T0204139

The Electronics Industry Research Series

- The Taiwan Electronics Industry
- The Singapore and Malaysia Electronics Industries
- The Korean Electronics Industry

THE Taiwan

ELECTRONICS INDUSTRY

Chung-Shing Lee
Michael Pecht

CRC Press
Boca Raton   New York

Acquiring Editor: *Norm Stanton*
Senior Project Editor: *Susan Fox*
Cover Design: *Denise Craig*
Prepress: *Gary Bennett*
Marketing Manager: *Susie Carlisle*
Direct Marketing Manager: *Becky McEldowney*

Library of Congress Cataloging-in-Publication Data

Lee, Chung-Shing .
 The Taiwan electronics industry / Chung-Shing Lee, Michael Pecht.
  p. cm. — (The electronics industry research series)
 Includes bibliographical references and index.
 ISBN 0-8493-3170-6 (alk. paper)
  1. Electronic industries — Taiwan. 2. Electronic industries —
Government policy — Taiwan. I. Pecht, Michael. II. Title.
III. Series.
 Hd9696.A3T355   1997
 338.4'7621381'0951249—dc21   97-147
                                                                                                  CIP

Contents

PREFACE

In the early 1970s, electronics was targeted by the government and people of Taiwan as a strategic technology and was heavily promoted through various policies. Since then, Taiwan's electronics industry has achieved outstanding success. At present, the electronics industry, especially the semiconductor and the information products sectors, is characterized by rapid growth and continued high potential.

This book investigates the past performance, current status, and future development of Taiwan's electronics industry. The information is based on the authors first-hand assessment of over 100 electronics companies in Taiwan, academic studies of Taiwan's electronics, computers, and semiconductor industries, and Chinese language sources including information obtained from Taiwan's Ministry of Economic Affairs, Institute of Industrial Technology Research, and Institute for the Information Industry.

This book is for readers interested in the historical development, current status, and future growth of Taiwan's electronics industry. Researchers and policy makers who need to know more about the role of central government in promoting "strategic" industries and in assisting national science and technology development in an Asian Newly Industrialized Country will find this book useful. This book also provides engineers with information about the current and potential capabilities of Taiwan's electronics technology. In addition, this book provides corporate business planners and managers in the electronics and other industries the information needed to comprehend the industry structure, general strategies, specific product market focus, and current performance of Taiwan's dynamic electronics industry. This information will be helpful for making decisions to form joint ventures and strategic alliances with Taiwanese manufacturers. Data about specific information products, communication products and electronics companies can be found in the Appendixes.

This book consists of nine chapters and begins with a brief overview of

Taiwan's geography, demographics and languages, government structure, educational policy, and labor force. (Chapter 1). Chapter 2 discusses Taiwan's present economic condition, public finance and industry status, the national goal of becoming the Asia-Pacific Regional Operations Center, as well as the general history of economic development and philosophy of economic policy. Government policies to promote the development of science and technology are addressed in Chapter 3, which includes an examination of the dynamic interrelationships among Taiwan's central government, research institutes, universities, and private industries, as well as the role of the science park in national technology development. The recent progress in building the National Information Infrastructure is also discussed in this chapter.

The subsequent chapters - the main body of the book - focus on the Taiwanese electronics industry. Chapter 4 presents the historical development of Taiwan's electronics industry, addressing issues such as government policy in assisting the formation and growth of the computer and semiconductor industries. In addition, the growing trends of Taiwanese firms in forming industrial groups and strategic technical alliances with foreign multinational corporations are examined. Chapter 5 surveys the current technical status and global market share of Taiwan's computers, communications, and consumer electronics segments of the electronics industry. Chapter 6 describes Taiwan's domestic technological infrastructure and the product status as well as market potential of the electronic parts and components industrial segment. Semiconductor products including microprocessors, chipsets, computer memory, application-specific integrated circuits (ASIC), discrete components, opto-electronics, and multiple chip modules (MCM). Current development of flat panel display technology is discussed in Chapter 7. Chapter 8 analyzes the current structure, strategy, and performance of Taiwan's electronics industry. A group of top 250 Taiwanese electronics companies was studied to evaluate the industry financial performance in 1995. This book concludes with an assessment of the current development and future directions of Taiwan's electronics industry (Chapter 9).

ACKNOWLEDGMENTS

This effort was supported in part through the MANTECH program grant No. 60NANB500060 administered out of NIST.

Many people have supplied us with technical information and other resource materials, and we are extremely grateful to them. These people include Dr. Linjun Wu of Taiwan's Institute of International Relations, Dr. Gang Shyy of the National Central University, Dr. Cher-Min Fong of the National Sun Yat-Sen University, George Lin of the CompuTech Services, Inc., James E. Beverly of MIS International, Fei-Fei Huang of the China Productivity Center, and Fang Ju Wu of the Industrial Technology and Research Institute who arranged use the library resources at ITRI.

We would also like to express our appreciation to Joanyuan Lee who provided research assistance and helped in the preparation of the manuscript, and Bill Huang for making some of the tables and figures.

Chung-Shing Lee is especially indebted to the George Washington University faculty members Dr. Lee Burke, Dr. Nicholas S. Vonortas, and academic mentor and friend, Dr. Lan Xue, for their teaching and advising. In addition, employer Dr. Edward J. Heiden, senior Vice President Dr. John Pisarkiewicz, Jr., and colleague Stephen C. McGonegal provided time to enable work on this book to proceed without interruption. Finally, Chung-Shing Lee dedicates this book to his parents, Wen-Chen Lee and Yun-Chu Lee, whose sacrifices and support made this book possible.

College Park, Maryland

AUTHORS

Chung-Shing Lee is a research economist at the Heiden Associates, Inc., a Washington economics and management consulting firm specializing in the application of economic and statistical analysis to business strategy and public policy. Mr. Lee graduated from National Taiwan University and the University of Maryland at College Park. He is also studying toward a doctorate degree in Engineering Management at the School of Engineering and Applied Science, the George Washington University. He has more than seven years of industrial consulting experience, including computer and telecommunication industries. His major research interests are in the areas of Asian technology development, strategy and technology management, industrial economics, and strategic use of information technology.

Michael Pecht is a Professor and Director of the CALCE Electronic Packaging Research Center (EPRC) at the University of Maryland. The CALCE EPRC is sponsored by over 40 organizations, and conducts research to support the development of competitive electronic products in timely manner. Dr. Pecht has a BS in Acoustics, a MS in Electrical Engineering and a MS and PhD in Engineering Mechanics from the University of Wisconsin. He is a Professional Engineer, an IEEE Fellow, an ASME Fellow and a Westinghouse Fellow. He is the chief editor of both the *IEEE Transaction on Reliability* and *Microelectronics and Reliability*, an associate editor of the *International Microelectronics Journal*, and on the advisory board of *the Journal of Electronics Manufacturing*. He serves on the board of advisors for various companies and consults for the U.S. government, providing expertise in strategic planning in the area of electronics.

Chapter 1

GENERAL INFORMATION

1.1. Geography

Taiwan is a small island province and the major territory under the control of the Republic of China. Located off the eastern coast of Asia in the Western Pacific, it is 377 km (226.2 miles) long and 142 km (85.2 miles) broad at its widest point. The island has a total area of nearly 36,000 sq. km. Taiwan is separated from the Chinese mainland by the Taiwan Straits, which are about 220 km (132 miles) at their widest point and 130 km (78 miles) at their narrowest. The island is almost equidistant from Shanghai and Hong Kong.

The fundamental topographic feature of Taiwan is the central range of high mountains running from the northeast corner to the southern tip of the island. Based on differences in elevation, relative relief character of rock formations, and the structural pattern, the island can be divided physiographically into the five major divisions: the Central Range -- a ridge of high mountains with a length of 270 km from north to south, volcanic mountains, foothills, terrace tablelands, and coastal plains and basins.

Taiwan is surrounded by warm ocean currents and enjoys an oceanic and subtropical monsoon climate noticeably influenced by its topography. The climate is subtropical in the north and tropical in the south. Except for the mountain areas, the mean monthly temperature in winter is above 15 degrees Centigrade (60 degree Fahrenheit).

1.2. Demographics and Language

According to census figures released in 1995 by the Ministry of the Interior, Taiwan's population numbered over 21 million as of January 1995. Taipei City has the highest population concentration (9,763 persons per sq: km), followed by Kaohsiung City (9,220 persons per sq. km.) in the south. In fact, about 59 percent of Taiwan's population is concentrated in four metropolises (Taipei, Kaohsiung, Taichung, and Tainan). Densely populated urban areas have

1

merged around Taipei, forming an interdependent economic and industrial network.

The population structure underwent great changes in the last few decades. The population distribution curve according to age groups indicates a baby boom in the post-war years between 1951 and 1955. The annual birth rate held steady over the next three decades, from 1956 to 1985. The end of the post-war baby boom was punctuated by the 1986-1989 interval, a period in which the implementation of family planning and other demographic measures by the Taiwanese government resulted in increased years of education, late marriages, the rise of nuclear families, and comparatively fewer potential mothers between the ages of 20 to 34 have reduced the birth rate. Since 1984, the population replacement rate has remained below one. It dropped to 0.8 in 1993. By 1994, the population growth rate had dropped to 0.87 percent, while the death rate was kept under 0.54 percent.

The national language in Taiwan is the same as the "Common Language" on the Chinese mainland (Mandarin). Most Taiwanese people also speak their own local language call "Taiwanese" or "Fukienese" and have a basic understanding of English because it is taught mandatorily as the major foreign language throughout high school.

1.3. Government Structure

Taiwan is a democratic society with a capitalism-oriented economy. The government comprises three main levels: central, provincial, and county/city, each of which has well defined powers. The central government consists of the Office of the President, the National Assembly, and five governing branches (called "yuan"): the Executive Yuan, the Legislative Yuan, the Judicial Yuan, the Examination Yuan, and the Control Yuan.

The president is the highest representative of the nation and is granted constitutional powers to conduct national affairs. The National Assembly and the Legislative Yuan jointly perform the functions of the parliament or congress of a western democracy. The leader of the Executive Yuan is usually referred to as the premier of the country. There are three levels of subordinate organizations under the Executive Yuan: the Executive Yuan Council; executive organizations, which include the eight ministries (Interior, Foreign Affairs, National Defense, Finance, Education, Justice, Economic Affairs, and Transportation and Communications) and several independent commissions; and subordinate departments, including the Directorate General of Budget, Accounting, and Statistics and other special commissions such as the National Science Council and the Council for Economic Planning & Development.

The Judicial Yuan is the highest judicial agency of the State. The subordinate organs of the Judicial Yuan are the Supreme Court, the high courts, the district courts, the Administrative Court, and the Committee on the Discipline of Public Functionaries. The Examination Yuan is responsible for the examination, employment, and management of all civil service personnel in

Taiwan. The Control Yuan is the highest control body of the State, exercising the powers of impeachment, censure, and audit.

The Taiwan Provincial Government exercises full jurisdiction over Taiwan's sixteen counties and all the cities except for Taipei and Kaohsiung. Taipei and Kaohsiung are special municipalities directly under the jurisdiction of the central government, rather than of the Taiwan Provincial Government. At the local level, under the Taiwan Provincial Government are five city governments - Keelung, Hsinchu, Taichung, Chiayi, and Tainan - and sixteen county governments, with their subordinate city governments.

Taiwan's Ministry of Economic Affairs oversees the nation's economic administration and development. The Industrial Development Bureau (IDB) of the Ministry of Economic Affairs (MOEA) (administrated by Executive Yuan) is the major governmental agency that deals with industrial affairs. The IDB initiates specific industrial policies and development strategies. It selects target industries and promotes development programs such as automation, industrial design, R&D, pollution control, and industrial parks.

The principal organization in Taiwan designed to facilitate closer cooperation between government and industry, as well as between Taiwan and its trading partners, is the China External Trade Development Council (CETRA), which is co-sponsored by the government and private industrial and business organizations. CETRA maintains forty-three branch offices, design centers, and trade centers in more than thirty countries. CETRA gathers trade information, conducts market research, promotes made-in-Taiwan products, organizes exhibitions, promotes product and packaging, offers convention sites, and trains business people.

1.4. Educational Policy

The government of Taiwan spends more money on education than on any other category except defense. Nine-year compulsory education was implemented in Taiwan beginning with the 1968 school year, significantly raising the general educational level of the people. The proportion of the population 6 years of age or older with a higher education was 1.4 percent in 1952, 3.7 percent in 1970, and 10.1 percent in 1988 [Taiwan Statistical Data Book, 1989]. In fiscal 1993, the education budget was US$ 15.5 billion, 40.5 percent of which went to elementary and junior high (compulsory) education, 6.4 percent to senior high education, 7 percent to vocational education, another 7 percent to junior college education, and 15.8 percent to university and college education. For fiscal 1994, 19.1 percent of the government's budget, or US$ 13.7 billion, was allotted for education.

There is little question that there has been very rapid educational growth in Taiwan, reflecting not only government policy but also a strong desire for education on the part of the people with deep-rooted traditional values. The Taiwanese view is that a strong education policy represents investment in human resources, and successful economic development requires an increasing number of people who are capable of doing useful research of industrial technology and who can apply research results to industrial production.

Over the years, the demand for higher education in Taiwan has been extraordinarily high. The perceived financial returns, gains in social status, and upward mobility associated with a diploma are reinforced by an important set of social and cultural factors. However, the heightened expectations associated with college diplomas create difficulties for the government's efforts to promote vocational schools and to meet the growing requirements of economic growth in the manufacturing areas.

1.5. Labor Force

Taiwan has a diversified and skilled national work force of roughly 9 million people with very low unemployment (national rate was 1.6 percent in 1994). Unfortunately, a characteristic of Taiwan's workforce is a shortage of labor, posed by a dwindling percentage of young laborers, and aggravated by a prolonged educational and two year military process. In fact, the percentage of workers between the ages of 15 and 24 plunged from 26.1 percent in 1981 to just 15 percent in 1994. Part of the government's solution to the labor shortage is to hire foreign nationals from countries such as the Philippines and Thailand. Unfortunately, the problem of labor shortage also caused a problem with illegal alien labor in Taiwan. The Employment Services Act declared in 1992 that aliens may not work in Taiwan without valid work permits.

In 1993, a greater number of Taiwanese graduate students began to return and the unemployment rate for people with college or graduate degrees was 2.2 percent in 1994. In 1995, the country witnessed a further increase in unemployment among those with a higher education. The rapid increase in the number of citizens seeking higher education has resulted in an increase of talented people where the market demand for their skills has not risen equally.

In 1994, 27.8 percent of the workforce was employed in the manufacturing sector, followed by commerce (21%) and social and personal services (14.4%). Agriculture and forestry remained relatively high (10.9%) compared to developed countries, and construction workers accounted for 10.8 percent.

The official minimum monthly wage in August 1995 was US$ 572 for a full-time worker. In 1994, the nationwide average monthly income was US$ 1,274. It was US$ 2,600 in the public utilities sector; US$ 2,000 in finance, insurance, and real estate; US$ 1,300 in both the construction and service sector; and less than US$ 1,200 in both the manufacturing and commerce sectors. Of the world's twenty-five most industrialized nations, the average hourly labor cost of Taiwanese workers in the manufacturing sector in 1994 was US$ 5.55, ranking 22nd. Germany ranked first with US$ 27.31, while Mexico was the lowest among the top nations with US$ 2.61. Asia's four little dragons ranked from 20th to 23rd, with Singapore (US$ 6.29) in the lead, followed by Korea (US$ 6.25), Taiwan, and Hong Kong (US$ 4.8).

Chapter 2

ECONOMIC DEVELOPMENT

2.1. Historical Overview of Economic Development

The island of Taiwan was occupied by Japan from 1895 to 1945. During that period, Taiwan was regarded as a colony that supplied raw materials and agricultural products to Japan and as a market for Japanese manufactured goods. There was never any intention on the part of Japan to industrialize Taiwan.

After the nationalist government retreated from mainland China to Taiwan in 1949, the government took over those enterprises established by the Japanese. In that period, Taiwan's economy was dominated by publicly owned enterprises; the statistics show that 56 percent of enterprises were publicly owned.

Since 1949, the pursuit of macroeconomic stability and low inflation have been the fundamental goals of Taiwan's policy (See Table 2.1). The economic development policy has consisted of five stages, in which the government has implemented comprehensive but changing policy packages.

- land reform and reconstruction (1949-52);
- import-substituting industrialization (1953-57);
- export promotion (1958-72);
- industrial consolidation and new export growth (1973-80);
- high technology and modernization (1981 - present).

In the 1950s, the industrial base in Taiwan was extremely weak, exporting principally only sugar, rice, and bananas. The per capita GNP was only US$ 145 in 1951, much less than that of many contemporary developing countries in Asia and Latin America. The import substitution policies adopted in the 1950s failed to sustain rapid economic growth in a limited local market. To react to this, the government adopted an export-promotion policy. Under this policy, not only was the sluggish economy revitalized, but the industrial structure in Taiwan was also changed.

5

Table 2.1 Economic and Political Timelines of Taiwan

Year				
1949		Nationalist Party takes power, Authorities nationalize Japanese assets	Land reform and reconstruction	Introduction of new Taiwan Dollar
1950	U.S. foreign aid			
1952				Land to Tiller Act
1953		1. High rates of interest on savings deposits 2. High tariff rates and non-tariff barriers 3. Multiple exchange rate	Import substitution	
1955				Exchange surrender certificates
1957				Low-interest export loans
1958			Export promotion	
1959				Nineteen-point program of economic and financial reform
1960				Statutes for encouragement of investment
1961		Third four-year economic development plan		Single uniform exchange rate
1964				Requirement for cement factories to export 100 percent of their production
1965				Statute for establishment and management of export processing zones
1968				
1969				National Science Council

Table 2.1 (Cont.)

1971				United Nations votes to expel Taiwan, China
1972			Launching of ten major public sector projects	
1973	1. Industrial consolidation 2. Direct public ownership and trade protection	Expansion of China productivity center with automation task force		Establishment of Industrial Technology Research Institute
1974				Adjustment Period
1975				
1976				
1980		Taiwan, China is forced out World Bank, International Monetary Fund		Establishment of Hsinchu Science-Based Industrial Park
1981	1. High-technology industrialization 2. R&D subsidies, brand development subsidies	Eighth Four-year development plan		
1982				Government targeted strategic industries
1984				
1985			Investment in 14 additional infrastructure projects	Appreciation of exchange rate
1986		Creation of opposition parties		
1987		1. Liberalization of foreign capital account 2. Establishment of standard labor laws	Lifting of martial law	
1989		1. Foreign exchange liberalization 2. Regulatory controls on bank loan rates and deposits are abolished		
1990				
1991				Sixth national development plan
1996				

Source: The East Asian Miracle, Economic Growth and Public Policy, A World Bank Policy Research Report, Oxford University Press, 1993.

By the year 1988, the per capita GNP was US$ 6,333, and in 1993 it reached US$ 10,566 -- the twenty-fifth   highest among countries with a population over one million, much higher than that of many countries that were ahead of Taiwan in the 1950s.

2.2. Philosophy of Economic Policy

Taiwan's industrialization was driven forward by the firm hand of an authoritarian development regime, while at the same time the country's development was hemmed in by high defense outlays and a scarcity of natural resources.

A good work ethics, the high propensity to save, the emphasis on education, the early adoption of an outward-looking development strategy, land reform, the crucial role of foreign aid in the early stages of development, and the favorable international economic environment have all, in various degrees, been among the important factors contributing to Taiwan's economic success. However, technological improvement, especially by the small and medium enterprises (SMEs), has played the most crucial role. In particular, nearly 54 percent of the growth in gross domestic product from 1952 to 1979 was due to technological improvement, while only 29 percent was due to labor growth and 18 percent to capital growth [Hou and Chang, 1981].

Taiwan's economy is structured for the small and medium-sized sectors, providing them with a recognizable high level of flexibility and dynamism. Taiwan promotes considerable foreign trade openness, since it has only a rather small domestic market involving about 20 million inhabitants. By resorting to a development and industrialization strategy targeting the installation of foreign productive capital and technology from the very outset, Taiwan has successfully realized a significantly higher per capita GSP (Gross Social Product) than Korea's, while having a generally symmetrical distribution of income and wealth. Counterbalancing the export success of Taiwanese companies on the global markets is the high market penetration by foreign companies in Taiwan. Currently, Taiwan is tightly tied into the global economy through direct investments.

In order to catch up quickly with the established industrial countries, the Taiwanese government has been playing a key role in the economy. Besides undertaking in steep investments in human capital, stimulating savings and capital formation, accelerating sectoral change, and supporting the export capability of the domestic economy, the Taiwan government also stepped forward as an entrepreneur, inducing the emergence of capital-intensive economic branches like energy production, petrochemicals, steel and ship building, aerospace industries and semiconductors. In comparison to Korea and Japan, Taiwan's industrial policy is heavily regulated and is only minimally affected by direct interventions in the market process (for instance, price and quantity controls or administrative directives). Such measures are highly transparent to both the domestic industry and third parties.

2.3. Current Economic Conditions

Taiwan's economy has undergone significant changes since the mid-1980s. Labor-intensive industries such as food processing, textiles, garments, and leather wares, which once dominated Taiwan's exports, have gradually been replaced by capital and technology-intensive fields such as chemicals, the information industry, and the electrical and electronic equipment industry. The percentage of export products by labor-intensive industries fell drastically, from 47.9 percent in 1987 to 38.8 percent in 1993, while the percentage of export products by capital- and technology-intensive industries rose from 22.4 percent to 32.1 percent during the same period.

In 1994, manufacturing output comprised nearly 80 percent of the total industrial production, or US$ 71.1 billion, and recorded a real growth rate of 5.7 percent. As of June 1994, some 2.5 million people were employed in the manufacturing sector, comprising nearly 30 percent of the national work force. Nevertheless, many manufacturers in labor-intensive industries have chosen to move their assembly lines to the Chinese mainland and other parts of the Asian-Pacific region to maintain their global competitiveness.

Overall, industrial outputs accounted for 37.3 percent (US$ 91 billion) of GDP, down from 1993's figure of 40.6 percent, due to a slowdown in industrial exports, while agriculture accounted for only 3.6 percent of GDP. The growing service sector reached 59.1 percent of GDP in 1994 due to stable growth in the financial, insurance, and commercial sectors. Private consumption, constituting over 57 percent of the GDP, increased by 8.2 percent in 1994, while government consumption shrank by 1.2 percent due to financial retrenchment and the five-year administrative cost-trimming plan. Although investment in the government sectors increased continuously in 1994, the growth rate dropped from 17.5 percent in 1993 to 11.2 percent. This was due to constraints in the government budget and the down scaling of the Six-Year National Development Plan.

In 1994, Taiwan became the world's twentieth largest economy and fourteenth largest trading nation. Its gross domestic product (GDP) grew 6.5 percent to US$ 244 billion, and its per capita GDP reached US$ 11,629. In 1995, Taiwan's GDP grew another 8.1 percnet ot US$ 264 billion, and its per captia GDP increased 7.3 percent to US$ 12, 439. Table 2.2 summarizes the performance of Taiwan's economy for the year 1994 and 1995. It is estimated that total GDP in 1996 will achieve US$ 278 billion, and per captia GDP will be over US$ 13,000.

Taiwan's trade with the rest of the world exceeded US$ 178 billion in 1994, and grew by 9.9 percent over 1993. The trade surplus in 1994 stood at US$ 7.7 billion, with a total export value of US$ 93 billion. The United States, Hong Kong, and Japan absorbed over 60 percent of Taiwan's exports. Major export items in 1994 included machinery and electrical equipment (40.6 percent of Taiwan's exports), followed by textile and textiles articles at US$ 14 billion, and telecommunications and transportation products (5.5 percent).

Table 2.2 Major Macroeconomic Indicators of Taiwan

Year	1994	1995
Gross domestic products (GDP)	US$ 2,439 billion	US$ 2,636 billion
GDP per capita	US$ 11,597	US$ 12,439
Foreign exchange reserve	US$ 92.5 billion	US$ 90.3 billion
Economic growth rate	6.54%	6.06%
Saving rate	26.61%	25.77%
Stock market index	6,253	5,544

Source: Bureau of Statistics, Executive Yuan, 1996.

Japan has been the largest importer to Taiwan, accounting for 29.2 percent (US$ 24.9 billion), followed by the United States 21.1 percent (US$ 18.1 billion). Major import items from Japan included machinery; electrical, electronic, chemical, and metal products, and auto parts. Agricultural and industrial raw materials accounted for 70.6 percent of U.S. imports, while capital equipment and consumer products made up 16 percent and 13.4 percent, respectively.

The value of Taiwan's trade with the Chinese mainland via Hong Kong surpassed US$ 17.9 billion in 1994, 18 percent higher than 1993. To diversify the country's outward investment, Taiwan implemented the Southern Investment Strategy, encouraging Taiwan businesses to invest US$2.4 billion in South East Asian countries in 1993, and as of 1994, investment totaled US$ 21.6 billion since the early 1980s. Growth in heavy industrial production has made up for the loss of labor-intensive industries that shifted production to South East Asian countries due to the appreciation of the Taiwanese (NT) dollar and high labor costs in Taiwan.

2.4. Public Spending and Financial System

Total expenditures at all levels of government was US$ 73.5 billion in 1994, an increase of 2.9 percent over the previous year. Among them, expenditures were US$ 44.6 billion in 1994, up 13.7 percent from 1993. Central government spending increased only slightly due to the five-year administrative cost-trimming program, but spending by local governments rose drastically because of higher spending on welfare.

Government spending on economic development in fiscal 1994 was 14.1 percent lower than 1993's US$21 billion. This decline was attributed to lower spending on the nearly-completed national development project. Expenditures on education, science, and culture accounted for 20 percent (US$ 1.5 billion) of total government spending. Social welfare expenditures rose by 12.9 percent from US$ 12 billion in 1993 to US$ 13.5 billion in 1994. Welfare spending is expected to be considerably higher in 1995 due to initiation of the National Health Insurance program and passage of the old-age stipend bill by the Legislative Yuan.

The combined revenue of all levels of government rose by 6.1 percent from US$ 54.2 billion in 1993 to US$ 57.7 billion in 1994. Of the total revenue, 58.5 percent came from tax and monopoly earnings, 17.6 percent from government borrowing, and 13 percent from state-owned enterprises. The government deficit fell slightly in 1994 to US$ 15.8 billion. The total accumulated government debt is expected to reach US$ 90 billion by the end of 1995 (36 percent of GNP).

Taiwan's high-growth economy, which lead to an increase financial activities over the last two decades, led to deep structural changes in the domestic financial markets. Increasing labor costs and appreciation of the Taiwanese (NT) dollar in the 1980s sped up the globalization of Taiwan capital by increasing financial and investment activities in overseas financial markets. This trend has applied competitive pressure on Taiwan's domestic financial system and invited foreign countries to open up reciprocal financial markets for foreign investors.

In fact, Taiwan's financial system has been under financial restructuring since 1987. Foreign exchange controls have been relaxed by removing restrictions on inward and outward capital flows. In addition, interest rates have been gradually liberalized since 1980. The current process of financial re-regulation and improved supervision in Taiwan has been in response to changes in the financial environment. The rapid expansion of financial markets, de-regulation of financial activities, financial innovations, new entries in the financial markets, and disorder caused during financial realignment all make re-regulation a necessary move.

Overall, the trend of the Taiwanese government's financial policy has been toward liberalization, which they see as the best way to make banking and finance more competitive.

2.5. Asian-Pacific Regional Operations

The Taiwanese government has declared its intention to develop Taiwan into an Asia-Pacific Regional Operations Center by promoting R&D targeting high value-added product development supporting capital intensive investment projects. The plan is to promote Taiwan as the location of choice for multinational enterprises wishing to set up headquarters from which to manage their operations in the Asia-Pacific area.

The Center is supported by six sub-operations centers, which include financial, manufacturing, sea transportation, air transportation, telecommunications, and media centers. The Council for Economic Planning and Development (CEPD) of the Executive Yuan is responsible for the overall planning and coordination of the project. The sub-operations centers are being handled by the Ministry of Finance, the Central Bank of China, the Ministry of Economic Affairs, the Ministry of Transportation and Communications, and the Government Information Office.

To advance the national project, the Coordination and Service Office for Asia-Pacific Regional Operations Center under the CEPD was set up in 1995 to serve international companies interested in taking advantage of the opportunities. Other specific measures for transforming Taiwan into a regional operations center include reducing import tariffs and further opening Taiwan's service sector to foreign companies.

Chapter 3

SCIENCE AND TECHNOLOGY DEVELOPMENT

It is often argued that government should play a significant role in the development of national scientific knowledge and technology, but government policy should not be a substitute for a market economy. Taiwan provides a good example of how government policy and market force complement each other to enhance science and technology development.

In general, direct scientific research in Taiwan is motivated first by profit - Taiwan's freewheeling market economy provides plenty of incentives for R&D in profitable technology; and second, by the National Science Council - the highest government office charged with coordinating national science and technology policy, which closely coordinates all R&D activities, and funds public-sector scientific and technological research projects through grants and subsidies.

3.1. Science and Technology Policy

The development of science and technology (S&T) in Taiwan can be divided into three stages (See Table 3.1). The foundation for science, education, and research was laid in the mid-1960s when US economic assistance to Taiwan was terminated and the country's new self-reliance focused attention to industrialization. A concurrent step was the strengthening of science education. In 1967 the National Science Council was established to initiate and coordinate science and technology development. In addition, the establishment of the Chung Shan Institute of Science and Technology and the Industrial Technology Research Institute (ITRI) solidified, the base for the development of the country's defense technology and applied industrial research respectively.

Following the first energy crisis in the early 1970s, the Taiwanese government seriously acknowledged the importance of promoting applied research. As a result, the Committee for Research and Development on Applied Technology was set up in 1976 under the Executive Yuan to oversee the coordinated promotion of applied science and technology by the relevant departments of the government. Also in this period, the Hsinchu Science-Based Industrial Park was established, along with various science and technology programs set up to promote national development.

Figure 3.1 illustrates the major R&D institutes established in Taiwan from 1979 to 1987. Since 1986, Taiwan has entered into the third stage of national S&T development. The third National Conference on Science and Technology was held, and the ten-year S&T development program has been implemented. It is estimated that by the year 2000, Taiwan will be able to join the group of developed countries.

The Six-Year National Development Plan was announced in 1991 as a major government policy designed to boost personal income, strengthen the development potential of local industries, balance regional development, and enhance the quality of life. It also targeted such problems as transportation congestion, chaotic urban development, air and water pollution, and inadequate cultural, medical, and recreational facilities. The plan aims to maintain an annual economic growth of 7 percent and to keep inflation under 3.5 percent for the next five years.

Table 3.1 Three Stages of Science and Technology (S&T) Development in Taiwan

Period Item	1st Period Basic 1966-1976	2nd Period Development 1976-1986	3rd Period Crucial time 1986-2000
Economic situation	- Improve environment for investment - Develop labor-intensive & export-oriented industries	- Readjust industrial structure - Encourage industrial research and development - Develop technology-intensive industries	- Join the group of developed countries
Emphasis on S&T Development	- Improve S&T education - Promote basic research	- Promote major trust programs and basic science - Training of S&T personnel	- Continue development in high technology

Table 3.1 (Cont.)

Major Measures	- Set up more departments and graduate schools at universities and improve quality of facilities - Establish five basic science research and engineering centers (1964-65) - Scientific Development Steering Committee (1967) - National Science Council (1967) - Chung-Shan Institute of Science & Technology (1969) - Telecommunication Research Institute (1968) - Industrial Technology Research Institute (1973)	- Committee for Scientific Development of the Executive Yuan established (1976) - 1st and 2nd National Conf. on S&T (1978, 1982) - S&T Development Program promulgated (1979, 1982) - Science and Technology Advisory Group set up (1979) - Establishment of Hsinchu Science Industrial Park (1980) - Academia Sinica formulated two five-year plans (1st: 1981-1986; 2nd: 1986-1991) - Training and recruiting of S&T Personnel program promulgated(1983) - SRCC established (1984)	- 3rd National Conf. on S&T held (1986) - Ten-year S&T Development Program (1986-1995) - Promotion of large- scale research projects - Encouragement of R&D in private industry
Economic Development GNP	US$2,361,132	US$1,132,841	GNP Target 2000 US$13,400

Source: K.T.LI, "The Role of Foreign Science Advisers in the Republic of China (Taiwan)" in W.T. Golden (ed.) Worldwide Science and Technology Advice, New York, NY: Pergamon Press, p. 138

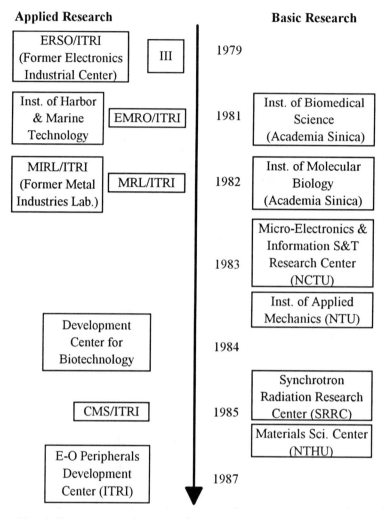

Abbreviations:
CMS: Center for Measurement Standards, ITRI
E-O: Electro-Optics
EMRO: Energy and Mining Research and Service Organization, ITRI
ERSO: Electronics Research and Service Organization, ITRI
III: Institute for Information Industry
ITRI: Industrial Technology Research Institute
MIRL: Materials Research Laboratories, ITRI
MRL: Material Research Laboratories, ITRI
NCTU: National Chaio Tung University
NTHU: National Tsing Hua University
NTU: National Taiwan University
SRRC: Synchrotron Radiation Research Center

Figure 3.1 Major R&D Institutes Established in Taiwan During 1979-1987.

The government is well aware of the fact that Taiwan has to accelerate its efforts in the development of science and technology in order to upgrade its economic structure, improve the quality of life of its people, and strengthen national defense. A ten-year S&T plan adopted for the period 1986-1995 targeted R&D expenditures as a percentage of GNP to increase from 1.04 percent in 1986 to 2.0 percent in 1995, the government's share in R&D to decrease from 60 percent to 40 percent in 1995, and the total number of researchers to increase from 1986's 27,747 to 43,000 in 1995.

3.2. R&D Expenditures

Taiwan spent a total of US$3.6 billion on R&D in 1992, an increase of nearly 16 percent over the previous year. R&D expenditures as a share of the GNP rose from 1.7 percent in 1991 to 1.79 percent in 1992. The public sector (all government offices and public enterprises) accounted for US$1.9 billion, or more than 52 percent of total R&D expenditures in 1992. Government offices accounted for over 45 percent of the total R&D expenditures in 1992; public enterprises accounted for 6.8 percent; private enterprises, 46.5 percent; non-profit groups, 0.8 percent; and foreign organizations, 0.5 percent. Conversely, the private sector accounted for US$1.7 billion, or nearly 48 percent of R&D expenditures in 1992. Over 98 percent of the private sector's R&D expenditures in 1992 was devoted to private R&D projects, up 0.6 percent from the year before.

As for R&D by sector of performance, research institutes spent US$ 1.2 billion on R&D in 1992, up nearly 24 percent over 1991. Colleges and universities spent US$500 million on R&D in 1992, 8.2 percent more than in 1991. Meanwhile, corporate spending on R&D rose 13.7 percent to US$1.9 billion in 1992. Large corporations have a tendency to invest more intensively in R&D than their smaller counterparts. Indeed, enterprises that spent US$ 400,000 or more on R&D accounted for 67 percent of all corporate R&D spending in 1992.

Considering types of research, spending on experimental development rose to US$ 1.9 billion in 1992, accounting for over 51 percent of all R&D expenditures in Taiwan that year. Conversely, spending on basic research totaled only US$ 454 million in 1992, or just over 12 percent of total annual R&D expenditures. In general, colleges and universities gave priority to basic research and applied research which accounted, respectively, for 43 percent and 48 percent of their R&D expenditures in 1992. Taiwan's National Science Council is encouraging institutions of higher learning in Taiwan to devote an even larger share of their resources to basic research. Corporate R&D expenditures were weighted heavily towards experimental development, which accounted for more than 70 percent of the total amount.

With respect to Taiwan's R&D manpower, over 77,750 people in Taiwan were working on R&D-related projects in 1992. Of these, more than 48,000 were researchers - that is, persons currently engaged in R&D activities, who hold B.S., M.S., Ph.D., or associate degrees, and have more than three years of research experience outside the classroom. The number of researchers rose by 4.7 percent in 1992. For every 10,000 people in Taiwan, 23.3 were researchers in 1992. The corporate sector employed over 26,000 researchers in 1992, up 4.9 percent over the previous year. Colleges and universities employed nearly 11,500 researchers, an increase of over 13 percent from 1991, while research institutes employed just under 10,750 researchers in 1992, a drop of 3.2 percent from 1991.

Researchers in Taiwan are assisted by more than 22,000 technicians, who perform technical tasks under the supervision of scientists and engineers. Technicians are typically graduates of high school, vocational school, or junior college, and have less than three years of work experience. Both the researchers and the technicians rely heavily on supporting personnel, nearly 7,300 of whom were employed at scientific or technical institutes in 1992.

3.3. Fiscal Policy to Promote S&T Development

The government has adopted a number of measures to encourage business firms to intensify their R&D efforts. On the taxation side, the Statute for Encouragement of Investment (SEI) stipulates that the R&D expenses of a firm must be deducted from the taxable income of the current year, and that accelerated depreciation is allowed for equipment bought for the purpose of conducting R&D with a service life exceeding two years. Furthermore, if the amount of R&D expenses of a firm in a tax year exceeds the highest amount of R&D expenditures in the five preceding years, then 20 percent of the excess amount may be deducted from the firm's income tax payable for that year; however, the deductible amount cannot exceed 50 percent of the firm's income tax payable for that year.

On the financial side, the government introduced the Assistance Program for Strategic Industries (APSI) in 1982. The government initially selected 151 products as strategic products for development; by December 1987, the number of products had increased to 214. Almost half of the selected items were electronic or information products. To implement this program, the government put aside NT$ 20 billion to give to firms in the form of loans. These loans have been used to install the machinery necessary for the production of strategic items. Any qualifying firm can receive a ten-year loan with a maximum loanable amount of 80 percent of the total capital needed, or 65 percent of the total expenditure required for the investment. Furthermore, a preferential interest rate, 1.75 percent below the prime rate of the Bank of Communications, is charged.

The SEI also specifically laid down minimum R&D levels for both domestic and FDI enterprises that enjoy substantial preferential tax treatment under the statute. Under the minimum R&D stipulations, if an enterprise's R&D expenditure to annual total sales ratio is lower than the prescribed standard, the enterprise is required to contribute the difference to a government-controlled

R&D fund for financing collective research and development projects. The required ratios vary according to the characteristics of the business and by the annual sales value of the enterprises; the range of the required ratios is from 0.5 to 1.5 percent. Enterprises in traditional sectors, such as the food and garment industries, have lower ratios than those of more technology-intensive industries.

Currently, the government is actively engaged in introducing a new Industrial Upgrading Statute (IUS) to replace the SEI. The major difference between IUS and SEI is that under IUS, firms will qualify to enjoy preferential treatment only if they have shown that they meet the prescribed standards in certain areas. This includes conducting R&D, compliance with environmental protection standards, engaging in manpower training, and setting up international marketing channels; under the SEI, a capital investment by either a newly established or an expanding firm is all that is necessary to enjoy tax-exempt preferential treatment. Clearly, under the IUS, the accumulation of capital is no longer considered the only factor in promoting industrial upgrading in Taiwan.

3.4. Government-sponsored Research Institutions and Universities

Structurally, Taiwan's economy was built on a large number of SMEs. Due to their limited resources for conducting R&D, SMEs rely heavily on the efforts of the government to develop technology and on government- sponsored research institutions to transfer technology to them. Consequently, the latter are critical to technology development and diffusion in Taiwan.

To help SMEs in the information industry to overcome the various problems associated with conducting R&D, training, and marketing, the government has set up two institutions to assist firms: the Industrial Technology Research Institute (ITRI) and the Institute for the Information Industry (III). III was established in 1979; its major mission includes introducing and developing software, assisting government agencies and public enterprises in their computerization projects, training and educating information professionals, supplying market and technological information related to the information industry, and promoting the development and use of computer-related technologies. Thus, the major function of III is to support activities that are currently undersupplied. More importantly, activities such as manpower training and the dissemination of market information can compensate for disadvantages resulting from the economies of scale that prohibit SMEs from conducting such activities on their own.

ITRI is the largest of Taiwan's independent research institutes, with headquarters in Hsinchu and branch offices throughout the Taiwan area and about 5,800 employees, 70 percent of whom are engineers or scientists. ITRI was set up in 1973 under the direction of the Ministry of Economic Affairs. As a national research institute, its primary mission is to develop industrial technologies; as a private entity, it emphasizes collateral benefits to local industry and society as a whole. For the most part, ITRI carries out long- and medium-term research. Most ITRI projects are sponsored by the government, and the research results are then transferred to the private sector.

ITRI generated US$ 520 million in revenue during 1994. Over 62 percent of its funding came from government-sponsored projects; 38 percent was received from the industrial sector for contract research, joint development, and technical services. In fiscal 1994, ITRI transferred new technology to 452 scientific and technological companies in Taiwan, hosted over 890 conferences and exhibits, and published 621 reference papers and 641 conference papers. Most indicative of ITRI's success was the number of patents it was awarded in fiscal 1994 -- a total of 368, of which 170 were foreign patents. In the same period, ITRI provided technical services to over 20,000 technology companies in Taiwan.

The III programs complement ITRI's hardware development program for the electronic/information industry in Taiwan. Five major divisions in ITRI support the development of various industrial technologies. One of these, the Electronic Research and Service Organization (ERSO), is critical to technology development for the electronics industry. ERSO has two major tasks: to develop needed technology for the industry and to diffuse the developed technology among the industry's firms. There are various mechanisms that can be used to diffuse the new technology, including issuing technical documents and organizing conferences for electronic firms in Taiwan; moreover, the new technology can be transferred to individual firms through licensing agreements which levy royalty charges on the recipient firm(s) once they employ the technology. If, however, the developed technology has marketing potential that might best be exploited by a new joint venture to disseminate the technology, then a new spin-off venture company could be established by ERSO. Note that although the spin-off company is supported by ERSO's engineers and its funding is provided by the government, the venture company is cautiously organized as a privately owned company. Both the technology endowment received from ERSO and the capital endowment supplied by the government aim at attracting investment from the private sector, which is expected to account for at least 60 to 70 percent of the total shares of the newly established venture company. The mechanisms that allow ERSO to diffuse its technology are summarized in Figure 3.2.

As a result of the successful spin-off model developed by ERSO, since the early 1980s, many integrated circuit (IC) companies began operation in Taiwan, such as United Microelectronics Corp. (UMC), Advanced, Taiwan Semiconductor Manufacturing Corp. (TSMC), and Winbond Electronics Corp. Figure 3.3 illustrates companies that were successfully spun-off from the ITRI organization during 1980s.

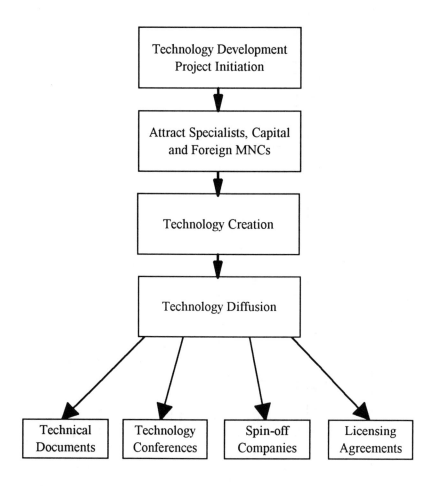

Figure 3.2 Mechanism for ERSO to Diffuse Its Technology

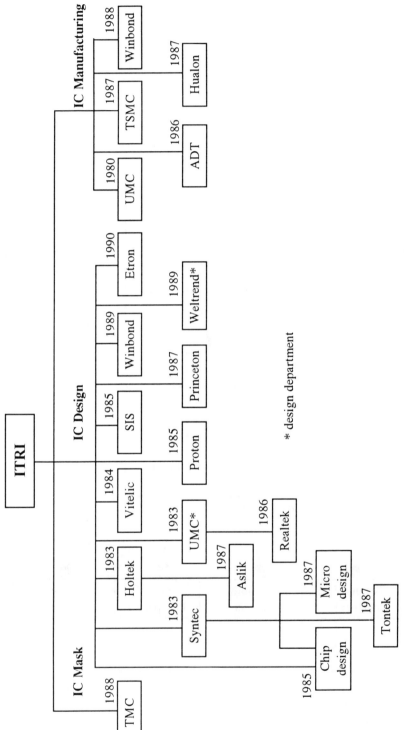

Figure 3.3 Companies That Were Successfully Spin-off From the ITRI Organization During 1980s

In addition to ITRI and III there are in the administrative structure of the government a number of institutes, councils, and the like that are charged with the responsibility of advancing science and technology in general and industrial technology in particular. Institutions that belong to the president's office include the Guidance Committee on the Development of Science and the Academia Sinica -- the highest academic institution in the country. Those under the Executive Yuan include the Committee of Advisors on Science and Technology; the Committee on R&D of Applied Technology; the National Science Council (NSC); the Council of Atomic Energy; the Council of Agricultural Development; the Ministry of Economic Affairs; the Ministry of Communications; the Ministry of Education; the Administration of Public Health; the Ministry of Defense; and the Taiwan Provincial Government (there are eighteen research institutes under its control doing research largely on agriculture, public health, and transportation).

The National Science Council plays the key role in the development of science and technology. It is charged with designing research strategy and plans, promoting basic research, pioneering applied research, and coordinating research undertaken by the various government agencies. It is also the principal grantor of funds to researchers at Academia Sinica and in Taiwan's colleges and universities.

Figure 3.4 summarizes the coordinated and cooperative relationships among government, universities, research institutes, and industries in the development of Taiwan's science and technology infrastructure. The division of research labor in technology development is presented in Table 3.2. As it indicates, basic research is conducted mainly by Academia Sinica and the universities, while applied research, technology development, and commercialization of the technology are mainly the responsibility of various industrial technology research institutions, as well as the enterprises themselves.

3.5. The Impact of Globalization

One of the important strategies used by Taiwanese firms to cope with the globalization of the world economy is overseas alliances or mergers. This strategy is common practice for many firms in Taiwan. Several significant economic factors contributed to the adoption of this strategy. Financially, huge foreign exchange reserves make overseas mergers feasible. Furthermore, with the appreciation of the Taiwanese (NT) dollar over the period of 1986 to 1988, overseas mergers and alliances have been made more affordable and attractive than exporting.

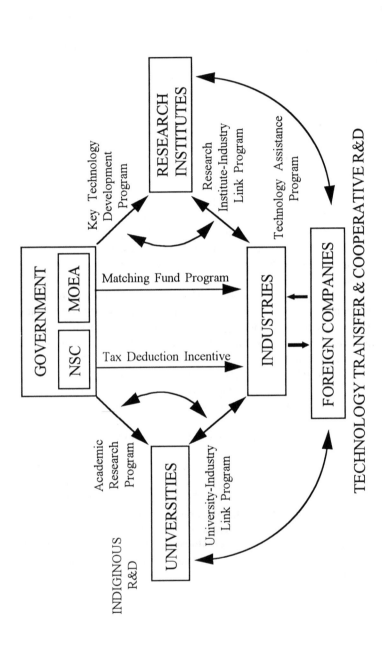

Figure 3.4 Relationship Between Government, Universities, Research Institutes, and Industries in the Development of Taiwan's Science and Technology Infrastructure

Table 3.2 The Division of Labor to Promote Technology Development in Taiwan

Type of research	Policy making	The implementation of the policies		
		Schools and research institutions	Government support research institutions	Enterprises
Basic applied research	Academic Sinica	Academic Sinica		
	Ministry of Education	Universities		
	National Science Council			
Technology development Commercialized and applied	Various government ministries	Various direct research institutions under the different levels of government	ITRI	Private enterprise
			III	Public-owned enterprises
			Other government-sponsored research institutions that are not directly under the government	

Source: National Science Council, Taiwan, R.O.C.

In addition to the financial factors, acquiring needed technology is another major motivation for overseas mergers. For example, Counterpoint Computers, a minicomputer manufacturing company, has helped Acer acquire minicomputer technology and expand its personal computer functions; and Microtex's overseas merger with the U.S.'s Mouse System Company enhanced its technology capability. In addition, many entrepreneurs in Taiwan find that overseas mergers and alliances are among the most feasible ways to compensate for the various disadvantages of domestic SMEs in the global markets, such as having a weak marketing network and little product brand-name recognition.

In sum, overseas mergers and alliances are appearing not only as a result of huge foreign exchange reserves and the vast appreciation of the Taiwanese (NT) dollar, but also because they help domestic enterprises to vertically integrate themselves and to acquire a foreign firm's marketing channels, brand names, and technology. This is a complementary and feasible strategy for domestic firms in coping with the globalization of the world's economy. With the experience they are developing in international legal and accounting services, Taiwanese firms' overseas mergers and alliances should dramatically increase.

3.6. The Role of Foreign Direct Investment in Technology Development

Taiwan's economy mainly consists of small and medium enterprises (SMEs). This situation directly affects technology development. First, because most of the SMEs in Taiwan were developed initially based on simple labor-intensive processing technologies, their technology levels are generally rather low. Second, most of the SMEs in Taiwan have devoted relatively little resources to R&D, due to their limited capital and manpower. One of the easiest ways for Taiwan's SMEs to develop technology is to become affiliated with a major foreign manufacturer, thus becoming their original equipment manufacturing (OEM) supplier. The SMEs becoming OEMs in Taiwan frequently form alliances with Japanese firms. In terms of foreign direct investment (FDI) in Taiwan, beginning in the early 1980s, Japan topped the U.S. and became the largest FDIs country investing in Taiwan.

From 1952 to 1987, electronics, metal products, chemical products, and machinery and instruments were the top four industries that attracted the majority of FDI in Taiwan. European firms average the largest investment, followed by U.S. and Japanese firms.

FDI enterprises in Taiwan can diffuse their technologies to domestic firms through various direct and indirect channels. One of the most typical channels for technology diffusion from the FDI to the domestic firms is labor mobility. Former FDI-firm employees working as managers or technicians believe that their experience made a significant contribution to strengthening management technology, product design, and marketing.

Besides FDI and technical cooperation with foreign firms, domestic firms can acquire technology from many other sources. A large-scale survey was conducted by the Directorate-General of Budget Accounting and Statistics of the Executive Yuan in Taiwan to study the major sources of technology of 4,226 manufacturing firms in 1985. Of these, 63 percent said that the firm's own R&D was their major source of technology, whereas 31 percent of them considered that their technology came mainly from abroad, by way of purchasing formulas and authorization or physical plants, foreign technological cooperation, foreign consultants, improving the products of other countries, and so on [Hou and San, 1994].

Among the three major domestic sources of technology (the firms' own R&D, purchasing domestic patents, and joint research with local research institutes), the in-house R&D in reverse engineering efforts was the most important source of technology. However, due to limited resources and diffusion channels, small firms relied more on their own R&D efforts, whereas medium or large firms used more joint projects with local research institutions to acquire needed technology.

The firms' own R&D efforts, assistance offered by local research institutions, technology cooperation with foreign partners, foreign consultants, and reverse engineering techniques (such as imitation or improvement of existing foreign products) are thus the five major sources of technology for manufacturing firms in Taiwan. Under such circumstances, the role that the government plays in technical advance is critical, for government policies are closely involved with R&D activities, government-sponsored research institutions, regulation of technology cooperation with foreign partners, and the protection of intellectual property rights.

3.7. Science Park and Technology Development

Overseas Chinese have played an important role in facilitating technology development in Taiwan. With this in mind, the Taiwanese government set up the Hsinchu Science-based Industrial park (HSIP) in 1980. By 1994, 165 firms with sales of US$ 6.7 billion were operating on the 580-hectare site. The administration of the park is a public undertaking supervised by the National Science Council (NSC). Unlike the tax and duty-free export-processing zones in Kaohsiung, Nantze, and Taichung, which were designed to attract foreign investment for export expansion and to transfer technology, HSIP was designed primarily to attract investment in high-tech industries, especially by returning students and overseas Chinese.

Newly established companies are eligible for a five-year income tax holiday. Older companies are eligible for a five-year tax holiday on income generated from expansion projects. In addition, a 20 percent tax credit is given to investors in park-based businesses that purchase additional shares in companies through rights issues [Bloomberg, 1995].   The proximity of National Chiao Tung University and National Tsing Hua University and the adjacent ITRI, not to mention convenient access to Taiwan's transportation network, have combined to give the HSIP a reputation as "Asia's Silicon Valley."

Hsinchu affords a favorable transportation hub, attractive living conditions (leisure and sports facilities, English-language schools, available apartment rental space), proximity to two technical universities, developed areas for industrial use (buildings and acreage for leasing) and an extensive menu of industrial services (transportation and loading facilities, warehousing, administrative expediting of imports and exports, disposal of industrial waste and effluent, information services). Currently planned is an expansion of the park's acreage by one-third; over the long term, Hsinchu is expected to form the core of a metropolis numbering 1.2 million inhabitants.

Firms based in the HSIP sold nearly US$4.8 billion worth of technology-intensive goods in 1993, up 41 percent from 1992. During the same period, companies producing integrated circuits in the park enjoyed sales of US$ 2 million in 1993, a substantial increase of over 73 percent over the previous year, signaling bright prospects for the future of Taiwan's IC industry. The total amount of business conducted by the park's electro-optics companies soared more than 47 percent to US$ 137 million, while that in the telecommunications sector rose over 8 percent to US$ 518 million. Makers of computers and peripherals, the park's largest sector, increased sales by 42 percent in the first eight months of 1993, having sold US$ 2.08 billion worth of technology-intensive goods during that period. As of June 1995, there are companies operating within the park, including IC (51), PC/peripherals (39), communication (30), opto-electronics (25), precision machinery (16), and biotechnology (9), employing nearly 36,000 workers. The park's revenues are expected to grow 60 percent, reaching US$ 10.3 billion in 1995. Total business turnover just broke the US$ 3.76 billion mark in 1993 [CNA, 1995].

The U.S. Silicon Valley experience has spawned numerous science parks around the world. Most productive cases of science park development combine an innovative milieu (synergy network), the capacity to reindustrialize on the basis of advanced competitive firms, and the ability to decentralize from traditional core locations into more dynamic start-up regions. Of the three major science parks in the Asian Pacific rim area - Japan's Tsukuba science city, South Korea's Daedok science town, and Taiwan's HSIP - HSIP so far is the most successful science park experience in Asia.

It is increasingly apparent that the establishment of the park has attracted many experienced and well-established Chinese scientists and engineers to return, especially from U.S. companies such as AT&T and IBM. As a result of this reverse trend, much advanced technology has been brought into domestic firms. Moreover, domestic firms in Taiwan now have more opportunities to internationalize their operations and the results of their R&D, further enhancing their global competitiveness.

The growth of the HSIP has run up against the realities of land acquisition in Taiwan. As a result, NSC has designated a site in southern Taiwan for a second science-based industrial park. The location is a 600-hectare park in Tainan County. Of this, 360 acres will be developed starting July 1996. The park will focus on serving companies from four major sectors: agriculture, biotechnology, micro-electronics, and precision machinery.

3.8. National Information Infrastructure

Due to the vital importance of information dissemination the national economic development, a "National Information Infrastructure (NII) Project Promotion Committee" was established in June 1994 to oversee the construction of the various national information and communication networks. Overall, NII is one of the objectives that must be accomplished in order to achieve the national goal of becoming the Asian-Pacific Regional Operations Center.

The NII committee consists of five sub-committees: resources planning (ruled by the CEPD), network construction (Ministry of Transportation and Communications), applied technology and promotion (Ministry of Economic Affairs), manpower development and basic applications (Ministry of Education), and general administration (Science and Technology Advisory Group). The government's role in this project is primarily to encourage investment and innovation through suitable tax and legislative measures. There is also a civilian consultative committee comprised of twenty-four members from leading private companies.

In conjunction with the promotion of NII and the provision of services to firms located in the Hsin-Chu Science-based Industrial Park for the use of R&D facilities, an integrated R&D information network has been established under the sponsorship of the National Science Council and the directorship of the National High-speed Computer Center. This new information network system has integrated the ten previously existing independent R&D information networks, such as the National Science Council, Science Park Administration, National Earthquake Engineering Research Center, IC Design and Manufacturing Center, Science and Technology information network, and High Precision Equipment Development Center, into a "National Science Council R&D Facility Service Information Network System."

The first concrete result from the project has been the "NII Experimental Hsin-Chu Broadband Network Region," which began operation in July 1995. A similar network began operation in Taipei during September 1995. In addition, the Directorate General of Telecommunications under the Ministry of Transportation and Communications is laying a nationwide fiber-optic cable network.

Chapter 4

THE DEVELOPMENT OF THE ELECTRONICS INDUSTRY

As Figure 4.1 indicates, the total output value of Taiwan's electronics industry in 1994 reached US$ 11.26 billion. The total value of the IC design and manufacturing output was US$ 1,231 million, almost 11 percent of the total electronics output value. This chapter begins with a discussion on the development of Taiwan's information electronics (computers and peripherals) segment of the industry in section 4.1. The formation, growth, and national strategy of the upstream electronic parts and components segment of the industry are examined in Section 4.2 and 4.3. Section 4.4 addresses the issue of the growing popularity of forming strategic alliances and industrial groups among domestic electronic products manufacturers and foreign MNCs.

4.1 The Growth of the Information Electronics Industry

The development of Taiwan's computer or information electronics industry can be categorized into four distinct stages [Hwang, 1995].

4.1.1. The Embryonic Stage (1978 to 1985)

During this stage, Taiwanese firms grew to utilize fully the industrial resources and experiences gained from manufacturing consumer electronics products. Government also played a significant role in planning and in directing resources into the computer industry. Major government involvements included establishing the Institute for Information Industry (1979); enacting the "information technology industry development plan" by the Economic Development Council of the Executive Yuan (1981); designating computer hardware system and system integration as the strategic technologies for future development (1982); passing the "strengthen training and acquisition of high-tech personnel act of 1983," which facilitated the return of the first wave of overseas Chinese scientists and engineers; investing NT$ 2.2 billion (US$ 63 million) (by the Executive Yuan, under the direction of the ITRI) to develop high-precision ICs (1984); and appropriating NT$ 6 billion (US$ 170 million) to facilitate and promote the development of high-tech key industries (1985).

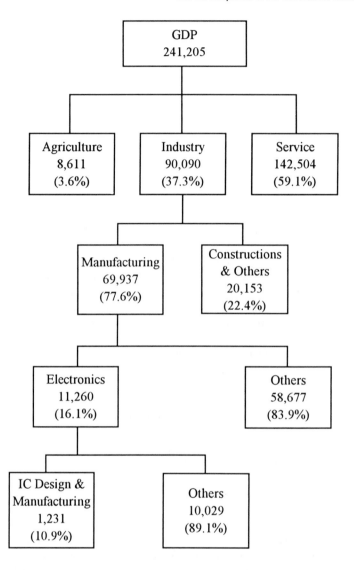

Unit: US$ million

Source: Taiwan ITIR, Project ITIS, 1995.

Figure 4.1 Estimated Contributions of Taiwan's Electronics Industry to the Overall GDP, 1994

The chief industrial activities in this stage included serving as OEMs of mainframe computer terminals and PC monitors, assembling computers for U.S. manufacturers, such as Apple computer, Inc. and manufacturing peripheral products such as keyboards, power supplies, and other computer components.

4.1.2. The Growth Stage (1986 to 1989)

Taiwan's division of industrial labor took shape in this stage. Computer manufacturers expanded their operations into several new areas, a step made possible by the large amount of capital and manpower injected into the industry. By the end of second stage, Taiwan's total annual production values in information technology reached US$ 5.48 billion. Several major events occurred in this stage: IBM decided to license its PC products to be manufactured in Taiwan, and Acer developed the first Intel-based 80386 32-bit microcomputer in Asia (1986); electronics products simultaneously became the top-ranked import and export categories, and Taiwan became the world's largest monitor maker, achieving more than 40 percent of the global market share (1987) and Acer, the largest computer maker in Taiwan, became publicly owned (1988).

In this period, Taiwan's computer industry expanded its overall industrial scale several times, established many new firms the motherboard and monitor industries, and enlarged domestic PC and component markets to other third world countries and brand-name markets.

4.1.3. The Turmoil Stage (1990 to 1992)

Through the early 1990s, Europe and North America suffered from economic recession. Sales growth of Taiwan's own brand-name PCS almost came to a halt. In 1992, global industry leaders, such as Compaq, AST, and DELL, began to reduce their product prices by 30 to 40 percent to overcome the slow sales. Further, the rising cost of labor in Taiwan and lower profits contributed to the failure of many domestic and foreign manufacturers; several internationally recognized firms (e.g., Zenith) ceased operation in Taiwan altogether. Even Acer had to layoff several hundred employees, delaying its brand-name marketing plan and returning to being an OEM in order to survive.

During this stage, industry was characterized by slow growth, companies adjusted their business strategy, global division of labor became important, and the production of foreign companies decreased rapidly, while domestic firms became the main force of the industry.

4.1.4. The Large-scale Investment Stage (1993 to 1996)

This stage was characterized by investment of domestic firms. Industrial structure grew increasingly concentrated and economy of scale became the key to success. In addition, upstream and downstream industries became highly integrated, and strategic alliances or joint ventures became ever more popular. A large volume of overseas contracts made Taiwan the world's major computer maker in this period. Taiwanese computer makers were able to accumulate a large amount of capital and use it to develop new technologies, investing in key electronic components (e.g., semiconductor and display devices), and constructing new production facilities. Newly formed industrial groups among major computer manufacturers also play a significant role in setting the direction of the industry. Finally, since computer, communication, and consumer electronics technologies (3Cs) have become highly integrated, domestic computer manufacturers have expanded into the communication equipment market.

The most significant event in this period was the operation of the National Information Infrastructure (NII) Experimental Hsin-Chu Broadband Network Region. Since then, the Taiwanese national information and communication basic infrastructure has formally entered into the experimental application stage. Future development and nationwide expansion of the NII will be based on the experiences of running this system.

4.2. The Development of the Semiconductor Industry

The prosperous growth of Taiwan's information electronics sector would not be possible without the successful development of the semiconductor segment of the electronics industry. Both industrial segments complement each other in overall electronics technology development. In this section, the focus shifts from the downstream computer-related products to the upstream electronic components and building the national technology infrastructure.

4.2.1. Historical Backgrounds

In the early 1970s, the government of Taiwan realized the need for one industry to lead the way in developing its high-technology export-oriented strategy. The electronics industry was selected for this role after detailed evaluations of the impact the move would have on the nation's economy. However, the electronics industry of Taiwan in the 1970s was still mainly an assembly industry. The key challenge for the government was to design and implement a strategic plan that would help the electronics industry develop into a technology-intensive, export-leading industry.

The single most important component for the electronics industry is the semiconductor. In general, the manufacturing of semiconductors comprises four separable phases: design and mask making, wafer fabrication, assembly, and final testing. Since the 1970s, a pattern of sub-regional specialization has emerged in the Asian-Pacific semiconductor industry. In Taiwan, indigenous firms were established to assemble semiconductors and compete with foreign multinational firms. These entrepreneurs are most often engineers and technicians, frequently overseas Chinese, who have gained experience working for foreign companies. Over a period of four years in the mid-1970s, around NT$ 410 million (a significant figure then) was invested in purchasing the required manufacturing technology, product design and testing, training, technical personnel, instruments, and equipment, and in building a pilot IC plant.

Since 1975, the government of Taiwan has played a leading role in the development of Taiwan's electronics industry, especially the semiconductor sector, and government policies have reflected this close relationship. Table 4.1 illustrates the milestones of Taiwan's IC industrial sector since the beginning of transistor packaging in 1966. Since then, the industry has achieved remarkable success in terms of increased sales volumes and profits, the growth in scale and scope of firms, a rising level of technological sophistication, and increased productivity.

Table 4.1 Milestones of Taiwan's IC Industry

Years	Milestone
1966	Kaoshiung Electronics established and dedicated to transistor packaging
1974	Taiwan Industrial Institute Electronics Center established Electronics Technology Advisory Committee formed by foreign experts
1976	Taiwan Industrial Technology Research Institute (ITRI) and RCA, USA, signed MOS IC technology corporation contract
1977	Electronic Center IC experimental factory was in operation
1978	First batch of digital watch IC was formally manufactured by Electronic Center
1979	Electronics Center renamed as Electronics Research and Services Organization (ERSO) ERSO successfully designed and developed multi-frequency remote decoding IC
1980	ERSO manufactured first batch of Bipolar ICs for the Taiwanese Market
1981	ERSO successfully invented touch tone telephone IC
1982	United Microelectronics Corp. (UMC) was formally in production

Table 4.1 (Cont.)

1984	ERSO developed a series of voice synthesis IC
1985	ERSO successfully developed 1.5 μm 256K CMOS DRAM
1986	UMC successfully developed 1.25 μm manufacturing process technology
1987	Taiwan Semiconductor Manufacturing Co. (TSMC) was formally in operation and had some experimental productions
1988	UMC successfully developed 1.0 μm 256K SRAM and 1M ROM
1989	UMC secondary factory started in production
1990	UMC manufactured 0.8 μm 1M SRAM
1991	ERSO started to build 8 inch μm laboratory
1992	ERSO μm laboratory utilized 0.7 μm to produce 256K SRAM 8 inch wafer
1993	ERSO μm laboratory experimentally produced 0.5 μm 16M DRAM
1994	ERSO μm laboratory regrouped as VANGUARD
1995	Vanguard, TI-Acer (1B), TSMC 3rd factory, and UMC 3rd factory completed 8-inch wafer fabs

Source: ERSO/ITRI ITIS project, 1995.

For analytical purpose, many researchers have attempted to divide certain specific stages to examine the developing process of Taiwan's semiconductor industry [Liu, 1993; Wu, 1993]. This chapter extends the previous analytical framework to study the development of Taiwan's semiconductor industry by categorizing it into five major stages: embryonic period, technology acquisition, technology creation and diffusion, growth stage, and technological collaboration. Each stage is discussed in the following sections. Overall, the Taiwanese government's efforts in promoting the semiconductor industry through various national projects since 1976 is summarized in Table 4.2.

Table 4.2 Taiwan's National Semiconductor Projects 1976-1995

	Development phase I	Development phase II	VLSI integration	Sub-micron
Time period	1976-1979	1979-1983	1983-1988	1990-1995
Official R&D outlays (mil)	NT$ 489	NT$ 796	NT$ 2,921	US$ 120
Goals and duties	Development of preliminary pilot production	Development and expansion of semiconductor infrastructure	Development of a production line, including production of masks	Development of technological competence in IC production in gauges down to 0.35μ
	Contract production for Taiwanese companies	Production of photo- masks	Development of CAD design capabilities	Cooperation of government ERSO laboratory with Taiwanese private industry (UMC,TSMC Etron, Holtek, MXIC, Mosel, Vitelic, Winbond)
	Advanced training	Learning latest process technologies	Development of technological competence in the VLSI field	
	Acquisition of technological competence in design and production of ICs	Development of competence in CAD design	Learning latest process technologies	
	technology transfer	Advanced training		
	Demonstration	Expansion of semiconductor technology to consumer industries		

Table 4.2 (Cont.)

Realized gauges	7.0 μ CMOS Complementary metal-oxide semiconductor	3.5μ CMOS	1.0μCMOS	0.35μ CMOS
Spin-offs		UMC (1980), Syntek, Holtek	TSMC (1987), Hualon (1987), Winbond (1988), Taiwan Mask Corporation (1988); various chip designers	

Source: Authors' own summary based on: Liu (1993), Howell, et al. (1993), ERSO; and on-site research.

4.2.2. Embryonic Period (1966 to 1976)

During the 1960s, when Taiwan's government realized that Taiwan would have to master the latest engineering technology, K.T. Lee and Tsin-Wan Zhao jointly organized the Modern Engineering Technology Symposium, which attracted engineers from home and abroad. The first symposium was held in 1966, the same year when General Instrument Microelectronics established its transistor packaging, which marked the beginning of Taiwan's semiconductor industry. At the meeting, the important role of the electronics industry was acknowledged by the participating scholars.

In the subsequent years, other MNCs, such as Phillips, Texas Instruments (TI), and RCA, also started semiconductor packaging operations. At the same time, several domestic firms, including Orient Semiconductor and Lingsen Precision Industries, also entered the business but all of them had only assembly capability.

In 1974, scholars and specialists at the Fifth Symposium agreed that Taiwan's electronics industry could not afford to remain in the assembly stage of production. Participants also came to the conclusion that Taiwan should not follow the step-by-step IC development common to more advanced countries, but would need to acquired appropriate technology as quickly as possible. To achieve this goal, specialists advocated bypassing the manufacture-of-transistors stage, directly manufacturing ICs.

Having obtained the government's support, organizers quickly established an advisory committee in the United States to coordinate with Taiwan's ITRI. Consequently, objectives and a manpower training program were jointly formulated. This advisory committee later became the Electronics Industry Research Center and later the Electronics Research and Service Organization (ERSO). ERSO's most important tasks were to secure the trust of the Ministry of Economic Affairs, to set up an IC demonstration factory, and to acquire technology from abroad for eventual transfer to the domestic private sector.

4.2.3. Technology Acquisition (1976-1979)

An agreement signed between the ITRI and the Ministry of Economic Affairs, effective in 1975, signaled the first official efforts to acquire appropriate technology. ITRI was thus the organization which in charge of acquiring generic technology and disseminating it to domestic firms. Subsequent developments in technology acquisition can be further divided into several distinct periods.

The most important tasks of the planning period were establishing the scope of the production technology, the focus of the acquired technology, the choice of a cooperating firm, and the nature of the cooperation. In 1976, the Taiwanese government selected RCA as the company to assist in transferring 7.0 μm CMOS (complementary metal-oxide semiconductor) process technology to domestic firms. During this period, an IC pilot plant, which included design, manufacturing, and testing capabilities, was completed with the assistance of RCA.

In the factory construction and manpower training phase (1976 through 1977), the major tasks included recruiting and training personnel, constructing the demonstration factory, purchasing and installing equipment, introducing design technology, and manufacturing standard cells. Finally, the goal of the production test period between December 1977 and June 1979 was to strengthen design and manufacturing ability, as well as to earn a high rate of return on investment in technology acquisition.

Overall, this stage was very successful. It provided a strong foundation to facilitate further technology transfer and development.

4.2.4. Technology Creation and Diffusion (1979-1983)

The major goals of this phase are to establish IC masking capability, the latest high-density process technology, and develop competence in computer-aided design (CAD). ITRI's technology capability advanced from 7.0 μm to 3.5 μm. In 1980 ERSO signed an agreement with Electromask of the United States to purchase MASK production equipment and accept manufacturing technology transfer. After the set up of the photo mask production equipment was completed in 1981, ITRI began to supply masks to domestic IC firms and to its own pilot plan. IC design technology was enhanced through the acquisition of computer simulation programs (e.g., SUSCAP II, SPICE II, and CICAP). ITRI then developed its own CAD software and mask design automation program. This greatly reduced the time needed to introduce new products.

Having accomplished this goal, the government adopted a technology diffusion strategy to assist in creating a domestic semiconductor industry. ERSO joined the Ministry of Economic Affairs to promote research and development in the electronics industry. The most important task of this stage was to transfer the technology acquired from RCA to enterprises in the private sector. To meet this need, ERSO adjusted its role, shifting its focus to MASK manufacturing and supply and design service, as well as to upgrading the capability of testing and applications.

In May 1980, ITRI spun off an entire IC manufacturing operation to create a new company, the United Microelectronics Corp. (UMC). It was Taiwan's first private IC manufacturer. Other spin-offs from ITRI in this period including Syntek Design Technology Co. and Holtek Microelectronics Inc.

4.2.5. Growth Stage (1983-1988)

Many ERSO personnel who had received several years of training in IC design were transferred to private industry or established their own businesses. ERSO, having cooperated with the private sector to develop IC design ability, continued to do research. In 1987, the Taiwan Semiconductor Manufacturing Co. (TSMC), with capital from the Taiwanese government (US$ 400 million), private investors, and Philips of Holland, was established to provide design houses with IC foundry services. This facility, with ITRI's assistance, acquired personnel and a US$ 250 million investment. The creation of TSMC allowed domestic IC design firms to operate without huge investments in manufacturing facilities.

By May 1978, TSMC had ninety-eight people with technological experience and forty-six production workers. This kind of valuable manpower resource enabled TSMC and UMC to avoid the high costs of recruiting and training personnel. This strategy stimulated the rapid growth in the number of independent IC design firms. Thus Taiwan's semiconductor industry experienced significant growth both in sales revenue and in technology capability during this period. Other spin-offs companies from ITRI in this stage included Hualon Microelectronics Corp, Winbond Electronics Corp., Proton Technology Inc., and other design houses.

Finally, in October 1988, ITRI spun off its mask-production department to establish a new company: Taiwan Mask Corporation (TMC). The structure of the industry was now complete, ranging from design, through masking, foundry, and packaging, to testing. During this transition, the role of the government changed from simply acquiring and transferring technology to cooperative research with the industry.

4.2.6. Government-private Technological Collaboration (1990-1995)

In 1989, Taiwan had six chip manufacturing companies and approximately fifty design houses. All aspects of the industry, including production, sales, design, and financial assistance, were established. During this stage, in order to cope with the rapidly changing technological environment in the electronics industry, the Taiwanese government has changed its industrial development policy. To overcome the scale disadvantage, government must not only maintain traditional technology acquisition and pioneer research, but also coordinate cooperative research and strategic alliances. Joint efforts similar to the U.S.'s SEMATECH consortium have been launched under the coordination of the Ministry of Economic Affairs in Taiwan.

The Submicron Process Technology Project was established in 1990. The alliance involved serval major Taiwanese IC manufacturers such as UMC, TSMC, Windbond, and Vitelic. In addition, Taiwanese government entrusting ERSO with U.S. $120 million to build a submicron laboratory, that could be rented by any firm. By 1994, Taiwan had 12 chip manufacturing companies, twenty-eight packaging firms, and over sixty-five design houses.

4.3. Summary of the Development of Taiwan's Semiconductor Industry

Compared with Korea's, Taiwan's semiconductor industry has much greater technological breadth, but so far it has not been able to achieve an absolute top position in any segment. Most of the some two hundred domestic semiconductor companies have specialized in chip design and software; somewhat more than ten companies produce integrated circuits.

Currently, Taiwan's production and exports are still focused primarily in the lower and mid-range technological categories. Only in 1993 was a breakthrough made in leading-edge technology. By 1994 five companies - TSMC, UMC, TI-ACER (a joint investment by Texas Instruments Inc. of the U.S. and Acer Inc.) Hualon and ERSO (Electronics Research and Service Organization - had established production facilities bases for 9-inch wafers and 0.5-0.3µ technologies; two additional companies (Winbond and Macronix) also plan to establish similar facilities.

On the product level, Taiwan's industry also achieved breakthroughs. Three companies - UMC, the U.S.'s Advanced Micro Devices Inc. (AMD), and Cyrix - are producing 486 SX microprocessors. Macronix is producing 32-bit RISC [reduced instruction set computer] central processing units. On the global scale, Taiwan's industry is most effective in the ASICs sector. Taiwan's competitiveness in this segment is based on the creativity and speed of numerous independent chip designers and the extensively developed division of labor in the production of ICs. (In contrast, the apparatus industry in Taiwan is stuck in the early stages. The demand for equipment is virtually covered exclusively through imports, half of them from the U.S. and half from Japan.)

The development of Taiwan's semiconductor industry can be accounted for beyond the comprehensive government support by the interaction of two factors: a technological push based on extensive investments by the U.S. semiconductor industry and a demand pull from Taiwan's computer industry. Direct U.S. investments since the seventies have led to the development of semiconductor production facilities locally and brought about a broad and intensive transfer of technology. It is estimated that some twenty percent of the engineers engaged in the semiconductor industry in Silicon Valley are of Chinese origin. Chinese returning to Taiwan have used their labor and management know-how and creativity to contribute decisively to the development of Taiwan's semiconductor industry. Taiwan's acknowledged strength in chip design and the software sector is also attributable in particular to this reverse brain drain.

A powerful impetus for growth arises from the demand of local consumer industries for semiconductors. This is especially true of the electronic data processing (EDP) industry, which consumes 70 percent of all semiconductors sold in Taiwan and 74 percent of all ICs. Because of successful sales of PCS (about one-tenth of global market production), notebooks, and ancillary peripherals (especially monitors, motherboards, scanners, networking equipment, mice, keyboards and terminals), Taiwan's EDP industry has developed into the fifth largest in the world. Its effectiveness is primarily based on the dynamism and flexibility of the mid-sized sector. Consequently, Taiwan's computer industry is likely to be well-equipped even for the looming turbulent era of multifunctional PCS and multimedia. Shifting wage-intensive production to the Chinese mainland has constituted a competitive advantage that is gaining increased significance; this has enabled the preservation of cost advantages and participation in China's expanding market.

In sum, Taiwan's IC industry developed with international assistance. Many of the initial proposals for developing an IC industry came from foreign scholars; technology was acquired from RCA during the early stages. A succession of foreign scholars returned to Taiwan to start businesses or to join others. This type of international cooperation enabled the industry to develop quickly. The government, although it played an important role in the industry's development, did not promote marketplace protection. As a result, Taiwan's IC industry has developed a high degree of competitiveness.

The most important reason for the industry's rapid development was the government's active involvement, which was adjusted as the industry matured. In the beginning, when the IC industry was still unfamiliar, the government established a demonstration factory, acquired technology, and trained personnel. Later the government, via ERSO, supplied manpower, technology, and management teams to direct the establishment of UMC and other design houses. Finally, the government invested capital, upgraded precision manufacturing technology, transferred equipment and managers to the private sector, and facilitated the creation of TSMC. This indirectly fostered a wave of private sector investment in the IC industry.

The success of Taiwan's IC industry proves that during the process of building this high-tech industry, the government was properly involved, lessening the business management risks and hastening development. On the other hand, the industry's development demanded certain requirements, such as manpower, technology, management ability, and capital. A key to success was the fine cooperation between government and the private sector.

4.4. The Formation of Strategic Alliances and Industrial Groups

Facing the emergence of intense global competition, growing market uncertainties, and rising constraints on resources, an increasing number of firms is seeking to form networks of national and international strategic alliances to achieve or maintain their competitive advantages.

A strategic alliance is a form of interfirm link. It combines specific aspects of the businesses of two or more firms that unite to pursue a set of agreed-upon goals while remaining independent. The partner firms share the benefits and control over the performance of assigned tasks. In addition, they contribute on a continuing basis to many key strategic areas such as technology [Yoshino and Rangan 1995]. Types of strategic alliance among worldwide electronics firms including non-traditional contractual agreements (R&D, manufacturing, marketing, and distribution/services, and so on) and equity arrangements (equity swaps and nonsubsidiary joint ventures). The proliferation of such collaborations can be attributed to three major factors: intense international competition, rapid technological advancement, and globalization [Culpan, 1993].

The successful growth of Taiwan's electronics industry can be attributed to the early technology transfer from industrialized countries, especially the U.S., and later, to intensive cooperative technology development efforts among private business enterprises, research institutes (e.g., ITRI and III), universities, public agencies, and foreign MNCs. The most famous case of technological cooperation is the formation of TSMC, a joint venture between Phillips and ITRI funded by Taiwanese government. TSMC not only accomplished the initial objective of supporting the domestic semiconductor industry's design capabilities, but also successfully established wafer manufacturing. Since then, the number of joint ventures and strategic alliances among Taiwan's domestic firms and foreign MNCs have grown exponentially. Figure 4.2 illustrates the formation of numerous strategic alliances between Taiwan's domestic IC industry and foreign MNCs as of 1995. Recent examples of strategic alliances with government participation in Taiwan's electronics industry are presented in Table 4.3.

Inter-firm alliances solely among private firms have also become popular in Taiwan. The alliance between Taiwan's two major computer makers, UMAX Data Systems, Inc. and Elitegroup Computer Systems Co. in 1993 provides an excellent example of intra-industry technological collaboration [Hwang, 1995].

UMAX is the top maker of images scanners in Taiwan; with an estimated global market share of 13 percent in 1995, it is second only to Hewlett-Packard. Elitegroup is one of the top three main circuit board makers in the world, and the most important overseas supplier for NEC of Japan. In addition, Elitegroup also manufactures CD-ROMs and monitors, and plans to introduce new multimedia-related products in the near future.

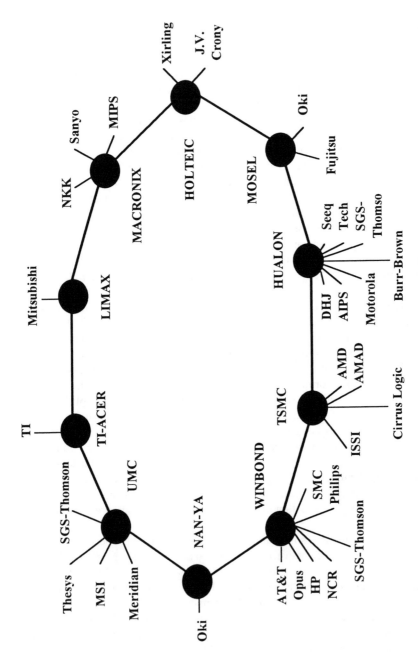

Figure 4.2. Formation of Strategic Alliances in Taiwan's IC industry (Source: ERSO/ITRI ITIS Project, 1995.)

Table 4.3 Examples of Strategic Alliance in Taiwan's Electronics Industry

	Notebook computer (information technology)	Laser FAX (consumer electronics)	Sub-micron (semiconductor)
No. of participants	46	11	9
Participated by government (agency)	Yes (ITRI)	Yes (ITRI)	Yes (ITRI)
Investment scale	small	medium	large
Public funding	20%	20%	50%
Close to commercialization	very close	close	distance
Scope of industry involvement	broad	medium	narrow
Value-added stage	product development	product development	process development
Problem Areas	- low product differentiation - intensive price competition - strong global competition - trade secret	- lack of key components - marketing - mass production	- public goods - capital investment - technological breakthrough
Period	July 1990 to March 1991	1990 to 1992	1991 to 1995

Source: ERSO/ITRI ITIS Project, 1995

The UMAX-Elitegroup alliance offers the partners the benefit of financial support from each other in new investments, while each partner retains flexibility and control in decision-making to cope with rapid market changes in its own main business. For example, Elitegroup has 8 percent equity in the Powerchip Semiconductor Co., which is a NT$ 20 billion (US$ 770 million) investment initiated by UMAX. This arrangement ensures a stable supply of semiconductor components. UMAX gets more control over product distribution and risk sharing in a large investment project.

In addition to inter-firm strategic alliances or joint ventures, other major Taiwanese firms are imitating Japan's operating strategy - that is vertically integrating all aspects of value-added activities from upstream component production to mid-stream assembly, to downstream marketing and distribution. For example, First International Computer Inc., which was established by the Formosa Plastics Corp. - the largest conglomerate in Taiwan, not only participated with Nan-Ya Technology (also established by Formosa Plastics Corp.) and Vanguard in a joint investment to manufacture 8-inch wafers, but also invested in other upstream activities, such as semiconductor testing and packaging businesses [Hwang, 1996]. On the other hand, gearing to the characteristics of each sales region, First International used a two-tier operating concept to develop marketing plans. In the Asian-Pacific market, First International developed its own distribution channels, based on its accumulated sales experience in Taiwan. On the other hand, in Europe and America, First International does not plan to set up its own distribution channels, so as not to create conflicts of interest with its OEM customers. On midstream activities, First International emphasizes product diversification, producing multimedia products, main circuit boards, notebook computers, image scanners, and fax machines. On the consumer electronics front, in addition to assisting NEC and Epson in manufacturing computers, First International is cooperating with AIWA of Japan to create a series of product development plans to introduce a new generation of multimedia computers. In the future, the First International group will consist of a cluster of information, communication, and consumer electronics technologies, able to link all value-added activities into an integrated system.

Chapter 5

CURRENT STATUS OF THE ELECTRONICS INDUSTRY

The electronics industry has become increasingly vital, especially for export-oriented, high-technology developing countries like Taiwan. In the 1990s, Taiwan's electronics industry experienced tremendous growth and prosperity in both product sales and technology development. Total electronics production value reached US$ 40.6 billion in 1995, a 58 percent increase since 1992. Tables 5.1 and 5.2 summarize the total production value and product trading statistics for Taiwan's electronics industry for the years 1992 through 1995. This chapter will consider Taiwan's electronics industry in terms of three sectors: computers (information electronics), communications and consumer electronics. In Taiwan these are called 3Cs.

5.1. Information Electronics

In 1995, information electronics was still the main force of Taiwan's electronics industry. By taking advantage of the fast growing global personal computer market and its own competitive advantage, Taiwan has achieved a 40 percent annual rate of growth. Total information electronics production value, which includes hardware, software, and overseas production, reached US$ 21.3 billion, a growth rate of 33.3 percent over 1994. This accomplishment has made Taiwan the world's third largest information product country, right after the United States and Japan. It also demonstrated that Taiwan's industrial competitiveness has significantly increased over other major global competitors.

Among the major information hardware products, monitors, personal computers (PCs), and main circuit boards are the three pillars of Taiwan's hardware industry. As Table 5.3 indicates, the global market share in 1995 was 57 percent for monitors, 10 percent for desktop PCs, 27 percent for portable PCs, and 65 percent for motherboards. In addition to leading in the production of monitors, PCs, and circuit boards, Taiwan has also become one of the major producers of image scanners, power supplies, graphics cards, and keyboards.

Table 5.1 Electronics Production Values by Category (Unit: US$ billion)

	1992 (%)	1993 (%)	1994 (%)	1995 (%)	1994-1995 growth rate
Consumer electronics	3.364 (13.1%)	2.710 (10.1%)	2.757 (9.0%)	2.767 (6.8%)	0.6%
Information electronics	9.292 (36.2%)	8.910 (33.4%)	10.254 (33.8%)	14.345 (35.4%)	40.2%
Communication electronics	1.793 (6.9%)	2.017 (7.6%)	2.007 (6.5%)	2.089 (5.2%)	6.3%
Electronics parts & components	11.230 (43.8%)	13.026 (48.9%)	16.395 (50.7%)	21.362 (52.7%)	39.1%
Total	25.664 (100%)	26.663 (100%)	30.376 (100%)	40.563 (100%)	33.8%

Note: Exchange rate 24.40 (1992), 26.63 (1993), 26.42 (1994), 26.48 (1995).
Source: Taiwan Industrial Production Statistics Monthly, ITRI ITIS project, 1996.

Table 5.2 Electronics Product Import-Export Statistics (Unit: US$ million)

	Export				Import			
	1992	1993	1994	1995	1992	1993	1994	1995
Consumer electronics	1,146	963	883	873	751	712	641	628
Information electronics	6,729	6,700	7,267	9,860	1,305	1,172	1,266	1,580
Communication electronics	798	1,072	1,095	1,241	468	378	313	503
Semiconductors	2,469	3,433	4,856	6,270	5,582	6,581	8,265	10,221
Electronics parts	8,894	10,254	12,166	15,390	3,686	3,982	4,378	5,883
Other electronics products	1,078	1,010	1,106	1,304	925	1,220	1,712	2,075
Total	21,117	23,432	27,373	34,938	12,719	14,045	16,574	20,890

Note: Exchange rate 24.40 (1992), 26.63 (1993), 26.42 (1994), 26.48 (1995).
Source: Taiwan Economic Development Council, ITRI ITIS project, 1996.

In terms of electronics hardware production value, Figure 5.1 indicates that Taiwan's domestic information production value reached US$ 14.2 billion in 1995, a 22 percent increase over the previous year. Furthermore, it accounted for 72 percent of Taiwan's total (domestic and overseas) information hardware production value. An analysis of Taiwan's current information industry for selected product categories follows.

US$ Billion

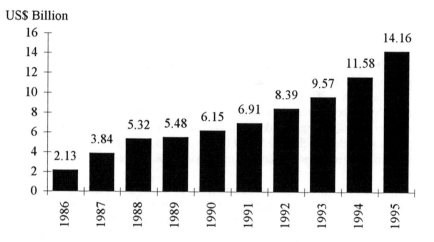

Source: Taiwan ITRI, 1996.

Figure 5.1 The Growth of Information Electronics Hardware Industry

5.1.1. Personal Computers

In 1995, Taiwan's sales volume of desktop personal computers exceeded 4.5 million units. This significant growth in Taiwan's desktop PC production and sales was due to further introduction of their own brand names by major Taiwanese manufacturers, and the large volume of OEM (original equipment manufacturer) orders from major foreign manufacturers. Overall, desktop computer production value and units manufactured enjoyed a healthy growth of 48 percent and 39 percent, respectively.

Taiwan was the third largest producer of notebook computers in the world in 1995 (27 percent of the world market share). Although sales of notebook computers in 1995 reached nearly 2.6 million units, 26 percent more than the previous year, several factors limited its growth. First, Taiwan's notebook computer industry relies heavily on the supply of LCDs from Japan, a potential threat to the future development of this industry. Second, there was a shortage of ICs components in domestic production. Third, technical problems (overheating) and manufacturing problems (volume production) have not been completely solved for high-end products, therefore, the growth rate in 1995 was not up to original expectations. Global market share has dropped 1 percentage point, to 27 percent.

Over the decades, the emphasis of the PC industry has gradually evolved from the highest value-added sector (relative to upstream components and software and to downstream brand name and marketing activities) to the lowest. The trend is expected to continue for the next four to five years. The strategy for PC makers is to expand the application area by integrating computers with consumer electronics and communications technologies.

5.1.2. Monitors

Due to a strong domestic industrial infrastructure, some 16 million monitors were produced by Taiwanese manufacturers in 1995. In addition, a healthy system of global division of labor, i.e., overseas production by Taiwanese-owned factories (mostly on the Chinese mainland, in Thailand, and in Malaysia) has produced another 15 million units, for a total of 31 million units of monitors. Total production value reached US$ 7.27 billion in 1995, giving Taiwan a 57 percent share in the international market and making it the biggest monitor supplier in the world.

Due to a continuous increase in labor costs and appreciation of the new Taiwan dollar (NT$), Taiwan has started transferring the production of low-end monitors to facilities in Southeast Asia and China. In 1995, Taiwan's domestic production units thus accounted for less than 52 percent of the total production. Nevertheless, its domestic production value is still increasing, due to the concentration on producing high-end products. Overall, the strategies for monitor manufacturers for the rest of the 1990s are first, to lower manufacturing costs by shifting production to Southeast Asia and mainland China; second, to close in on the major markets by investing in Mexico and Europe; and finally, to build up a global marketing network in order to obtain large OEM contracts.

5.1.3. Main Circuit Boards (Motherboards)

Taiwan's main circuit board industry has grown substantially over the past decade, primarily as a result of increasing OEM orders from American, European, and Japanese personal computer manufacturers, such as Compaq, NEC, Canon, and Fujitsu. In addition, the dramatic increase in sales of Pentium-based PCs and the introduction of Taiwanese brand name products also contributed to strong demand for Taiwan's motherboards.

The biggest event within the motherboard industry in 1995 was the entry into the market by Intel Corporation. It has been estimated that approximately 6 million motherboards were manufactured by Intel during 1995. Obviously, global competition has become more intense because of Intel's entrance into the market. Although Taiwan still remained the top supplier of motherboards in the world, its global market share dropped from 1994's 80 percent to 1995's 65 percent.

Nonetheless, strong demand in the global motherboard market has helped Taiwanese manufacturers maintain stable growth. Total 1995 production, which includes overseas Taiwanese-owned factories, was 20.9 million motherboards, with a total production value of US$ 2.46 billion, a 22 percent increase over 1994.

Table 5.3 Taiwan Major Information Electronics Hardware Product Statistics (Unit: 1,000)

Eelectronics hardware products	1994			1995		
	Domestic production unit	Overall production unit	World market share	Domestic production unit	Overall production unit	World market share
Monitor	14,391	24,028	56%	16,085	31,329	57%
Portable PC	2,057	2,057	28%	2,592	2,592	27%
Desktop PC	3,090	3,090	8%	4,167	4,567	10%
PC motherboard	11,529	17,545	80%	13,113	20,846	65%
Scanner	1,663	1,663	61%	2,481	2,481	64%
Power supply	12,150	25,960	31%	7,756	34,320	35%
Graphics card	5,040	8,770	32%	4,920	9,300	32%
Terminal	1,060	1,060	22%	956	956	27%
Network adapter	6,100	6,120	34%	9,946	10,264	38%
Mouse	22,052	29,800	80%	31,087	40,904	72%
Keyboard	7,068	22,800	52%	4,589	32,780	65%
Sound card	1,986	1,986	11%	3,188	3,750	22%
Network hub	310	310	18%	922	933	22%
Display card	472	472	24%	1,663	1,663	35%
CD-ROM	186	186	1%	2,825	3,752	11%

Source:Taiwan Institute for Information Industry, 1996.

5.1.4. Network Adapters and Image Scanners

In 1995 Taiwan's sales volume of network adapters reached 10.3 million units, 68 percent more than the sales volume in 1994 (see Table 5.3). For the next several years, the growth in total production value may suffer due to the maturity of technology and the escalation of price competition. Nevertheless, the increasing popularity of multimedia and network applications could boost the sales of network-related electronics products.

Meanwhile, in 1995, Taiwan's production volume of image scanners reached 2.5 million units, 50 percent more than the volume in 1994 (Table 5.3). The increasing availability of image processing application software and the decrease in hardware prices have contributed to the demand for scanners and to significant growth in production.

5.1.5. Miscellaneous

Taiwan also led the world in the production of keyboards, mice, and power supplies. In 1995, except for mice suffered a loss of 8 percentage points in global market share, power supplies and keyboard both experienced high rates of growth and increased worldwide market share.

Although the growth of Taiwan's CD-ROM drive production volume and total value were very limited in 1994 due to the late start of mass production, within just one year the CD-ROM sector had achieved remarkable results in 1995. As Table 5.3 indicates, overall production has increased from 1994's less than two hundred thousands units to 1995's nearly 3.8 million units. Global market share also increased by 10 percentage points from one percent to 11 percent. Today, more and more domestic manufacturers are producing CD-ROM drives.

As multimedia applications become ever more popular, the demands for CD-ROM products will also increase; therefore, the potential growth of this sector is very promising. It is expected that CD-ROM drives, along with other multimedia products such as MPEG chipsets and sound cards, will become another major information electronics products of Taiwan.

5.1.6. Information Electronics Industry Outlook

Overall, the outlook for Taiwan's information industry remains optimistic. The substantial development of the information industry and its product market reflect the continuous advance of information technology, the increasing speed of product release, a shorter product life-cycle, a substantial growth in demand from various types of users, and the integration of information technology into everyday activities. However, threats from the increasingly intensive competition among industry rivals, changing global economic conditions, and the continuous drop in hardware prices could cause a decline in production value in the near future. The immediate task for Taiwan's information industry is to continually and rapidly introduce new high-value-added quality products and services to satisfy users' demands. There are four major trends within Taiwan's information electronics industry. The impacts of these trends on the future development of the industry will be strong and long lasting. These four trends are:

- Developing design capability and brand name manufacturing. Many Taiwanese information electronics firms have gradually moved from being purely OEMs to develop design capabilities, and ultimately, to introduce their own brand names. As design capabilities and marketing skills of many Taiwanese electronics manufacturers become mature, more and more of them will gradually change from being product designers and manufacturers to brand name producers. The two largest portable computer makers in Taiwan, Inventec and GVC, have achieved 100 percent design capability selling products under their customers' brand names. Furthermore, Acer has already become one of the well-known brand name producers in the global market. In addition to Acer Corp., several other Taiwanese information electronics manufacturers have started to establish their own images in the global marketplace.

- Increasing activities in alliance formation and vertical integration. Taiwan's information electronics industry has become more centralized over the years, which has contributed to an increase in the corporate size, production scale, product quality, and credibility of Taiwanese firms. Mergers, acquisitions, and alliances among domestic manufacturers and product market diversification is growing. Larger corporations with diversified product and market strategies will dominate the smaller ownership/partnership single-product companies.

- Participating in global value-added activities. Since the early 1990s, many global information electronics firms, such as IBM, Digital, and Compaq, have gradually transformed their global production and marketing strategies. When configuring their worldwide value-added activities in a new competitive environment, global firms only keep the high-value added portions of the value chain, while outsourcing the rest of the value-added activities to other firms. This strategy has enabled many global firms to increase the speed of introducing new products, and at the same time, helped them to reduce cost, risk, and direct competitions from emerging firms. Many Taiwanese firms have taken advantage of this strategy to expand their design and manufacturing capabilities, as well as their global presences. For example, Compaq cooperated with Taiwanese firm Mitac in 1995 to reconfigure and coordinate the entire global value-added activities. Mitac took over the responsibilities of product design, manufacturing, procurement, logistics, and inventory activities, while Compaq kept only the brand name and distribution channel for itself.

- Changing nature of overseas manufacturing. Since 1995, many top-ranked Taiwanese information electronics firms, such as Acer Computers, Acer Peripherals, Lite-On Technology, GVC, and Liton Electronics, have increased the speed and amount of overseas investments. However, two major characteristics have distinguished it from the traditional overseas investment. First, the preferred locations of overseas manufacturing have gradually shifted from Southeast Asian countries to Philippine, Mexico, Chinese mainland, and Scotland [World Journal, 1996b]. Second, the outward investment model has upgraded from the traditional indigenous factories (local model) to cost-oriented and market-oriented factories (global model). The major purpose of this new strategy is to simultaneously achieve low-cost and close-to-market manufacturing on a global scale.

The growing significance of overseas manufacturing of information electronics products by Taiwanese companies is revealed by the increasing ratio of overseas to total industry production. In 1995, the value of overseas production already accounted for 29 percent (US$ 5.7 billion) of the total production value by all Taiwanese information electronics firms. It is expected that the trend and magnitude of global outsourcing, international division of labor, cross-country alliance activities, and the introduction of own brand names by Taiwanese manufacturers will continue. By the turn of the century, according to Taiwan's Institute for Information Industry (III), overseas information electronics output will increase to US$ 23 billion, or 46 percent of the industry's total annual production.

5.2. Communications

The communications equipment industry has been designated by the Taiwanese government as one of the ten emerging industries. The industry is most likely to follow information electronics and semiconductors, becoming Taiwan's next main industrial force.

5.2.1. Production Values and Market Focus

In 1996, there were about 284 communications-related equipment manufacturers in Taiwan. Of these, 55 percent are small and medium-size enterprises (SMEs); each has a total capital value less than US$ 1.9 million. According to the 1995 census of the communications industry conducted by ITRI, Taiwan's total domestic communications industry production value is US$ 2.73 billion, 15 percent more than 1994's US$ 2.38 billion (see Figure 5.2). In addition, overseas Taiwanese-owned factories added another US$ 526 million in 1995 to the overall production value.

Table 5.4 illustrates the export values of Taiwan's communications equipment by product category. The percentage of export value to total production value has increased tremendously from 1992's 48 percent to 1995's 76 percent. This trend has significantly reflected the export-oriented nature of the industry.

Geographically, major export areas in 1995 included North America (27%), Western Europe (25%), mainland China (17%), and Southeast Asia (12%). In addition, sales strategy for individual products were related to the specific market area. For example, all bureau exchangers and more than half of the cable communications equipment (58%) which were exported to the Chinese mainland and Southeast Asia were under the manufacturers' own brand names. On the other hand, cable CPE, wireless equipment, and network products, which targeted sales in the North America and Western Europe, were mostly OEM/ODM products.

As for product sales performance in 1995, except for bureau exchangers, faxes, and pagers had experienced negative growth, other communications products all demonstrated positive rates of growth, especially for ISDN CPE, network products, and satellite LNB, due to global Internet fever.

5.2.2. Communications Industry Outlook

The performance of Taiwan's communications industry in 1994 was below industry expectations due to intensive price competition in cable communications equipment, which accounts for more than one-third of Taiwan's communications equipment production volume; wireless communications equipment manufacturers are still suffering from a lack of technical breakthroughs to introduce new products. Nevertheless, Taiwan's production of communications network products has been flourishing due to the increasing sales of personal computers and the reduction in price of PC network adapters; at the same time, manufacturers of exchange machines and transmission equipment have opened new market channels in areas such as Southeast Asia and mainland China. This implies that Taiwan's capabilities in manufacturing communications equipment have become well recognized.

US$ million

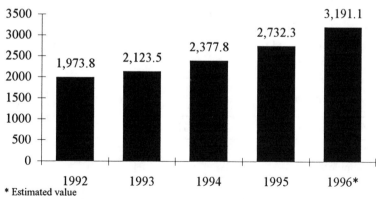

Exchange Rate: 24.40 (1992), 26.63 (1993), 26.42 (1994), 26.48 (1995-96).

Source: Taiwan ITRI, 1996.

Figure 5.2 Domestic Production Value of Taiwan's Communication Equipment Industry

Table 5.4 1994 Taiwan Domestic Communication Products Ranking (Unit: US$ million)

1994 Ranking	Product	1993 Export value	1994 Export value	1993-1994 Growth rate (%)
1	Bureau exchanger	665	566	-14
2	Modem	222	358	61
3	Telephone	272	354	30
4	Network product	260	341	31
	Network adaptor	203	250	23
5	Fax	147	207	41
6	Pager	80	98	22
7	Satellite LNB	63	95	50
8	KTS/PBX	67	83	23
9	DAML	-	45	-
10	Wireless communication equipment	49	41	-17
Others	Other communication product	305	178	-41
	Total	2134	2370	11

Source: Taiwan ITRI ITIS project, 1995.

Table 5. Estimated Growth of Taiwan's Communication Equipment Products (Unit: US$ million)

Product	1995	1996	1995/1996 Growth rate
Bureau exchanger	431.7	402.2	-7%
Cable communication equipment	140.5	143.5	2%
Cable CPE	1,294.6	1,668.8	29%
Wireless equipment	384.9	411.6	7%
Network product	435.1	565.0	30%

Exchange rate: 26.48, Source: Taiwan ITRI, 1996

In 1995, by taking advantage of strong global demands, the industry achieved 15 percent growth over the previous year. Figure 5.2 illustrates the steady growth of domestic production value since 1992. In the future, the global Internet fever will add to the already strong worldwide demand for communications-related products. Other communications products, such as direct broadcast satellite (DBS) TV and wireless communications, are also expected to benefit from the Internet fever. It is estimated that Taiwan's 1996 domestic production value of communications equipment could reach US$ 3.2 billion, 17 percent higher than in 1995. In addition, the overseas production value is estimated to be US$ 657.1 million in 1996, 25 percent more than 1995's value of US$ 526.1 million.

As for individual communications products, bureau exchangers and cable communications equipment are estimated to have very low or even negative growth due to the increasing saturation of the domestic market. On the other hand, Table 5.5 indicates three major communications products that are expected to have higher growth rates ranging from 7 percent to 30 percent. In addition, domestic production of fax machines, KTS, and network products is also expected to increase OEM/ODM contracts.

Overall, the global trend of telecommunications industry deregulation, combined with the advance of communications and computer technologies, will create even higher potentials for future business in this area. Four major characteristics of today's global communications market are pertinent to Taiwan's communications industry:

- The annual growth rate of the traditional telephone network is about 10% (an estimated 60 million units annually). Even though Asia is the fastest growing market in the world, the global market has been dominated by a few major international players, due to the maturity of their technologies and the high entry barriers to the market. On the other hand, voice-recognition technologies have become low-end products in end-user markets. Therefore, developing countries for which lower production cost is the comparative advantage will easily become the leading manufacturers of these products.

- Wireless communications has become well accepted as an end-user communication service. Over the last several years, the prices of wireless products and service charges have dropped significantly, due to advancements in electronics technology, improvements in the semiconductor manufacturing process, and economies of scale in production. In addition, major international players in this field are working toward establishing international standards and networks, that will provide individual end-users with wireless capability throughout the world using only one equipment standard. In response to this trend, Taiwan's communications industry is developing systematic design capabilities rather than confining itself to single-product development.

- Broadband communications was not expected to be available until the next century. However, due to the expansion of cable television networks, the increasing popularity of end-user computer network communications, and enormous efforts by governments to assist the development of broadband communications services, these services may well be available before the end of the century. Since the government of Taiwan began to participate in National Information Infrastructure (NII) projects in 1994, a new communications environment incorporating broadband concepts has started. It will provide not only new product development opportunities for Taiwan's domestic manufacturers, but also new communications services to end-users.

Due to the global fever of Internet on-line services and the strong demand for high-speed data communications, the Taiwanese government has decided to open high-speed digital switching communications services to both domestic and foreign private businesses [World Journal, 1996a]. Before the end of 1997, Internet services providers or other communications firms will be able to invest in their own facilities to expand services without restriction. It is expected that communications equipment manufacturers and end-users will both benefit from the new competitive market environment.

5.3. Consumer Electronics

Consumer electronics products once played a vital role in Taiwan's electronics industry, accounting for some 20 to 30 percent of total production value and exports. Surging information electronics, accompanied by the stiff appreciation of the NT dollar and a steep jump in labor costs, have hurt the local consumer electronics industry, inducing local manufacturers to invest overseas. The consumer electronics industry has been mired in a slump since 1990. In 1989, over 300 factories were operating in Taiwan; only 245 manufacturers still operate four years later, achieving a production value of US$ 2.8 billion in 1994 and 1995, down 18 percent from 1992 (see Table 5.1). In 1995, Taiwan's export value of consumer electronics was US$ 873 million, a mere one-percent decrease from the previous year. On the other hand, the total import value was US$ 628 million, a two percent decrease from the previous year (see Table 5.2).

While consumer electronics output constituted over 42 percent of the total export value in 1982, by 1994 the figure had plunged to below 5 percent. Since traditional products are still the staples of Taiwan's consumer electronics industry, the decrease in the export and import value of consumer products is not surprising. Developing countries in Southeast Asia and China are becoming major manufacturers of consumer electronics. Even advancing countries such as Japan, Europe, and the United States are facing the same competition as Taiwan. These countries respond by transferring production facilities overseas and concentrating on developing advanced products. The new generation of consumer electronics products will integrate 3Cs technologies to become intelligent and user-friendly personal or household products. The technological trend is toward digitalization and network. Two category of products seem promising in the near future:

advanced TV [e.g., high-definition TV (HDTV), digital TV, and widescreen TV (WTV)] and set-top box (STB) (e.g., Broadcasting STB and Internet TV). Recently, Taiwan has been successful in developing some new advanced consumer electronics products.

5.3.1. Advanced TV

Taiwan has been developing and promoting WTV since 1993. The sales volume of WTV in 1995 was approximately 15,000 units. In 1995, Taiwan's WTV was capable of digital signal processing. The main models include 28-inch and 32-inch WTVs. Proton Electronic Industrial Co., Ltd. has already introduced a 36-inch WTV capable of integrating communications and computer equipment. Although Taiwan's WTV market has demonstrated a promising future, the 16:9 CRT -- the major component of WTV, which accounts for about 40% of the total costs -- is still produced solely by Japanese manufacturers. Taiwan's Chunghwa Picture Tubes Ltd. is planning to manufacture the 16:9 CRT in the near future. This strategic move could reduce production costs and increase the competitiveness of Taiwan's WTV products. Table 5.6 compares the current level of HDTV and WTV technologies in Taiwan to those advanced technologies in developed countries.

5.3.2. Digital Video CDS

Japanese manufacturers have been strongly promoting the concept of digital video CD, offering over 400 titles in 1993. Taiwan's manufacturers were able to provide over 600 titles in 1994; Taiwan's TXC Corporation and Sampo Corporation have begun to manufacture video CD products in 1995. On the other hand, Taiwan's ability to manufacture CD-ROMs is still elementary, though its CD-ROM products have experienced a tremendous growth in 1995 that will substantially contribute to the development of the digital Video CD products.

5.3.3. Internet TV Receiver

The Internet is the most popular on-line service today, with an estimated of 56 million households worldwide already connected to it. However, most of them are experienced computer users. In order to reach the mass majority of customers who have little computer skills but also want to explore cyberspace, several companies, including Apple Computer Inc., have aggressively developed and introduced many versions of low-price, user-friendly "Internet appliances." In introducing these new type of products, companies were either adding communications capability to traditional consumer electronics or developing the Internet TV STB, which is capable of directly accessing on-line services. The recent status and future R&D direction of Taiwan's technological capability in this area are shown in Table 5.7.

Table 5.6 Current Level of Technology in Taiwan's HDTV and Wide-Screen TV.

Technology item	Development stage	Current level of technology in Taiwan	Advanced technology
Picture tube	R&D	29 inches	-
	Product	21 inches	40 inches/16:9
TV signal processor	R&D	Scan conversion	Frame conversion
	Product	Picture-in-picture	Picture-in-picture
Digital vision decoder	R&D	MPEG-2	MPEG-2
	Product	MPEG-1	MPEG-1/-2
Digital voice decoder	R&D	MPEG-2 multi-channel	-
	Product	MPEG-1 single channel	MPEG-2/AC-3
Digital transport decoder	R&D	16/64-QAM	256-QAM
	Product	-	64-QAM
Digital tuner	R&D	Double mode	-
	Product	Single mode	Double mode

Source: Taiwan ITRI, 1996.

5.3.4. Consumer Electronics Industry Outlook

The industry trend in the second half of the 1990s is toward the integration of computer and communication technologies with traditional consumer electronics. Many Taiwanese computer makers, such as Acer, First International, and Mitac International, have advanced into the consumer electronics product area with some success. In early 1996, several of Japanese global consumer electronics giants, such as Sony, Panasonic, Matsushita, and Hitachi, came to Taiwan to look for OEM partners for manufacturing a new generation of consumer electronics products. If alliances were established between Taiwanese computer makers and Japanese consumer electronics giants, Japanese firms would be able to provide alliance partners with well established brand names, product images, and good quality. On the other hand, Taiwanese firms could provide first class computer manufacturing capabilities to the partnership. The cooperation between the two countries and industries could strengthen the global competitive advantage of Japanese consumer electronics firms, at the same time, it could also further expand the growth prospects for the Taiwanese computer makers.

Table 5.7 Current Level of Technology in Taiwan's Internet TV Receiver

Technology item	Development stage	Current level of technology in Taiwan	Advanced technology
Web browser	R&D	TV version	HTML 3.0
	Product	PC version	HTML 2.0 plus
Java	R&D	In progress	V1.1
	Product	None	Java V1.0
Use model	R&D	TV version	-
	Product	PC version	-
Operating system	R&D	In progress	Oracle, Javasoft
	Product	None	-
Progressive monitor	R&D	In progress	-
	Product	-	-

Source: Taiwan ITRI, 1996.

Chapter 6

SEMICONDUCTOR MARKET FOCUS

6.1 Overview

Taiwan's semiconductor market is primarily an integrated circuit (IC) market. Development over the years is illustrated in Figure 6.1. The scale of Taiwan's IC market in 1995 is estimated to be approximately US$ 8,014 million -- a 40 percent growth over the previous year. Before 1990, Taiwan's IC industry doubled its capacity every four years. In recent years, the time required to double industry capacity has shortened; since 1991, Taiwan's IC industry has achieved this feat in three years. The major reason for this fast growth is the booming electronics industry, which has become a high IC-usage business.

Figure 6.2 displays Taiwan's IC market structure in terms of 1995 values. The total market value of US$ 8 billion in Taiwan's market came from two major sources: domestic ICs (US$ 1.45 billion) and import ICs (US$ 6.56 billion); both increased significantly over the previous year. Total product value increased from 1994's US$ 2.2 billion to 1995's US$ 3.7 billion, while import market value also increased in 1995 from US$ 4.8 billion to US$ 6.56 billion. Throughout the early 1990s, domestic ICs as a percentage of the total IC sales market in Taiwan remained close to 15 percent. In 1995, the percentage increased to 18 percent.

The application markets for ICs in Taiwan are concentrated mostly in information-related products; the total production value of Taiwan's information electronics industry increased 33 percent from 1994 to 1995. In fact, information-related ICs and DRAM/SRAM memory products represented 80 percent of the domestic IC application market. Other applications, such as modems and fax machines, also became the high growth areas with communications ICs accounted for 8.8 percent of the domestic ICs market. In addition, pagers and mobile phones also exhibited high potential for growth. Finally, the consumer electronics industry continues high demand for ICs, around 9.4 percent of the application market in 1995.

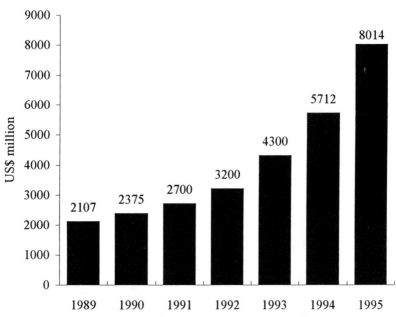

Source: Taiwan ITRI, project ITIS, 1996
Figure 6.1 The Growth of Taiwan's IC Market

Overall, U.S. and Japanese imports of IC products accounted for half of Taiwan's total IC supply in 1995. Figure 6.3 indicates that the U.S. currently is the largest supply source of ICs, providing close to 30 percent of the domestic market. This includes such products as microprocessors and peripherals. Japan accounts for some 20 percent of the domestic market supply, and its relative percentage is still rising.

Table 6.1 presents Taiwan's IC import areas between 1989 and 1994. The U.S. and Korea both increased their shares of import ICs over the period, while the shares of Japan, Hong Kong, and Europe declined in 1994. As for IC exports, Taiwan's five-year average compound growth rate was about 24.9 percent between 1989 and 1994. Table 6.2 shows the changing relative share of exports by geographic areas by Taiwan's IC manufacturers. The relative shares remained roughly the same over the five-year period, with the U.S. still on top, followed by Hong Kong and Southeast Asia. Southeast Asia has the highest rate of growth, due to strong demands for IC packaging and products from Taiwanese electronics firms. IC exports to Hong Kong are transferred almost entirely to mainland China.

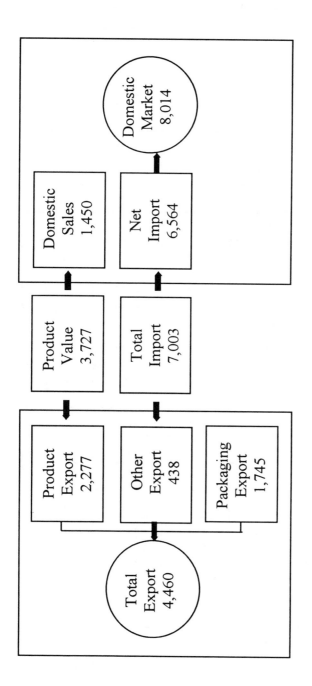

Unit: US$ million
Exchange rate: US$ 1 = NT$ 26.48
Source: Taiwan ITRI, Project ITIS, 1996.

Figure 6. 2 Taiwan's IC Industry in 1995

Table 6.1 Taiwan IC Import by Area and Year

Year Area	1989	1991	1993	1994
Japan (%)	39.6	26.4	26.4	23.4
USA (%)	30.0	35.0	30.0	32.0
South Est Asia (%)	10.4	18.9	16.0	15.0
Korea (%)	8.2	8.2	8.5	9.7
Hong Kong (%)	5.6	5.6	6.0	4.1
Europe (%)	2.8	3.1	12.4	11.6
Others (%)	3.4	2.9	2.3	4.2
Total (US$ million)	1,853	2,305	3,856	5,063

Source: Taiwan ITRI, ITIS project, 1995.

Table 6.2 Taiwan IC Export by Area and Year

Year Area	1989	1991	1993	1994
Japan (%)	8.5	9.8	6.6	6.7
USA (%)	34.0	24.9	34.9	31.6
South East Asia (%)	9.1	4.7	12.6	18.1
Korea (%)	0.2	8.2	4.4	4.6
Hong Kong (%)	13.7	15.9	21.3	18.2
Europe (%)	20.8	15.8	17.1	7.9
Holland (%)	12.7	19.6	12.0	11.0
Others (%)	1.0	1.1	1.2	1.7
Total (US$ million)	1,114	1,366	2,393	3,443

Source: Taiwan ITRI, ITIS project, 1995.

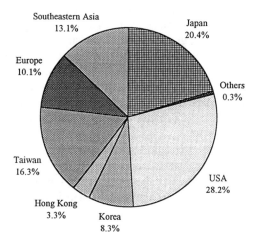

Figure 6.3 Supply of Taiwan's IC Industry by Area

6.2. Infrastructure

In 1994, the worldwide growth rate of IC production value was less than in the previous year. Nevertheless, Taiwan's IC domestic industrial production value achieved a remarkable 53 percent rate of growth. The current industrial framework of Taiwan's IC industry has gradually become an integrated industry. In contrast to the U.S. and Japan, which adopted a strategy of integrated operations, Taiwan's IC industry to a large extent employed a division-of-labor approach.

Currently, Taiwan's IC industry system consists of hundreds of individual highly specialized companies. Among the 120 Taiwanese IC firms in 1995, 55 percent of them are classified as IC design companies. The total number of IC design firms in Taiwan is still growing at a high rate, due to the small scale of operations, limited investment required to establish a new facility, and relatively low entry barriers.

The number of firms categorized as IC manufacturers is also growing rapidly in Taiwan. The total number of current IC manufacturers has reached 19 including Mosel-Vitelic, which began operations in the second half of 1994. In addition, a joint venture between China Steels and MEMC (an American firm), the Taisil Electronics Material Corp., is constructing an 8-inch wafer factory that will be ready to join the group of IC manufacturers in early 1997. In response to the increasing demand for IC products, other IC-related manufacturing activities, such as IC packaging, testing, conductor manufacturing, and chemical processing, have also expanded aggressively since 1994.

Overall, Taiwanese IC industry production value reached US$ 3.7 billion in 1995, as Figure 6.4 indicated. This is a 70 percent increase over the previous year. Between 1991 and 1995, the total value of domestic IC production has doubled in every two years, with a compounded average growth rate of 48.3 percent. Such high rates of growth in the IC industry mainly reflect the changes of emphasis in product development by domestic firms; that is, the development of information IC products has gradually shifted away from peripheral IC devices toward core products. In addition, increasing capabilities in the multiple design of memory products and the improvement in manufacturing processes also have contributed to the high growth rates. Finally, constant support from various peripheral industries has played a significant role.

On the domestic side, the factory expansion activities by Vanguard, coupled with the completion of UMC and TSMC's fab3, will both improve the condition of Taiwan's DRAM supply and also relieve the problem of capacity under-utilization (foundry services) encountered by domestic IC design companies.

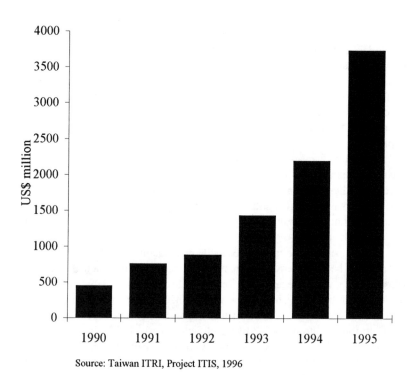

Source: Taiwan ITRI, Project ITIS, 1996

Figure 6.4 1990 -1995 Total Taiwan Domestic IC Production Value

6.2.1. Design

In 1995, Taiwan's IC design industry experienced two major events: First, the introduction of Pentium Triton chip sets by Intel in February has caused Taiwanese IC design firms to adjust their product line structures. Second, the worldwide price of SRAM dropped significantly during the fourth quarter of the year, which in term has reduced many IC memory design firms' revenues. Nevertheless, Taiwan's IC design industry still managed to achieve a remarkable 56 percent rate of growth in product sales.

Figure 6.5 demonstrates the scale of Taiwan's IC design industry since 1987. Major indicators of the scope of Taiwan's IC design industry are presented in Table 6.3; there were sixty-six IC design firms in 1995, with total sales of US$ 729 million. Table 6.4 displays Taiwan's top ten IC design companies and their sales and growth rates. The top three firms - Silicon Integrated Systems Corp. (SIS), ACER, and VIA - are also the major domestic suppliers of PC chipsets. Among the top ten firms, Sun Plus, operates as several technical teams, concentrates its R&D efforts on consumer-related IC design. Etron's specialty is high-speed SRAM design, and with the current of strong demand for cache SRAM, it posted a remarkable growth rate of 163 percent in 1994 and 105 percent in 1995.

With the start-up of Mosel-Vitelic's DRAM and SRAM productions, the need for Japanese foundry and packaging services was reduced dramatically in 1994.

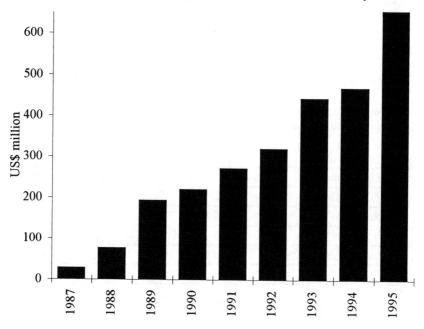

Source: Taiwan ITRI, project ITIS, 1996

Figure 6.5 Market Scale of Taiwan IC Design Industry

Taiwan IC design industry's reliance on domestic wafer fabs and IC packaging has since increased dramatically to 96.9 percent and 99.8 percent, respectively.

Table 6.5 indicates that the percentage of information-related IC design increased sharply, from 1993's 30.7 percent to 1995's 54.5 percent, due to the increasing demand for PC chipsets and VGA controller products. Because of Mosel-Vitelic's withdrawal from the IC design business, the percentage of memory IC design dropped to 22.5 percent in 1994 and 13.9 percent in 1995, while consumer-related IC design has a higher potential for growth due to expanding markets in mainland China and Southeast Asia. Finally, except for network adapter cards, the communication IC design business has the lowest growth rate, due to the limited expansion capability of the downstream industries.

6.2.2. Manufacturing

In 1995, the high rate of growth in Taiwan's IC manufacturing industry was attributable to the entry of Vanguard (an impressive US$196 million sales in the first year of operation) and the triple-digit sales growth of Winbond and Mosel-Vitelic. As a result, total industry sales reached US$ 4.5 billion, an increase of 70.4% over the previous year (see Table 6.6). Due to intensive competition, Taiwanese IC manufacturers have been trying to develop better manufacturing processes. At the end of 1995, major Taiwanese IC manufacturers (e.g., UMC and TSMC) were able to use the 0.3μm manufacturing process technology, an important milestone for Taiwan's IC manufacturing capability. The 0.25 μm process technology is expected to be developed and used in 1997.

Figure 6.6 displays the scale of Taiwan's IC manufacturing industry since 1987. Table 6.7 shows the 1995 ranking of manufacturers, total sales, and rates of growth of Taiwan's IC manufacturing companies. TSMC is ranked number one for sales two years in a row because of increased production capacity from new

Table 6. 3 Taiwan IC Design Major Indicators (Professional Design Company)

Year Item	1991	1992	1993	1994	1995
No. of companies*	57	59	64	65	66
Sales (US$ million)	270	319	442	468	729
Growth rate (%)	24	30	36	6	56
Import : export	49:51	50:50	46:54	65:35	61:39
Investment/sales (%)	71	10	23.5	15.54	15.9
R&D/sales (%)	9.9	10.1	9.5	10.0	12.2
R&D/manpower (%)	32	25	51	51	49
Average R&D experience (year)	4.5	4.7	5.4	5.9	5.0

* Including system manufacturers and design department of multinational corporations.
Source: Taiwan ITRI, project ITIS, 1996.

Table 6.4 Taiwan's IC Design Companies

Company	Total sales (US$ million)		1994~1995 Growth rate (%)
	1994	1995	
SIS	98.4	156.0	58.8
VIA	49.2	92.8	89.0
Acer	68.1	85.0	25.0
Sun Plus	39.4	55.9	42.3
Etron	26.9	54.9	104.6
Elan	-	43.5	-
Realtek	34.8	43.4	25.0
Princeton	11.0	25.3	131.0
Weltrend	12.5	18.9	51.5
Taiwan Memory	-	18.5	-

Note: Exchange rate 26.42 (94), 26.48 (95)
Source: Taiwan ITRI, project ITIS, 1996.

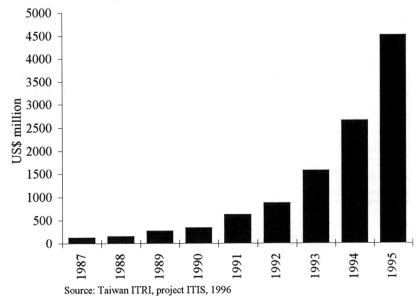

Source: Taiwan ITRI, project ITIS, 1996

Figure 6.6 Market Scale of Taiwan's IC Manufacturing Industry

Table 6.5 Taiwan IC Design Industry by Type of Application (Unit: %)

Application \ Year	Information	Memory	Communication	Consumer	Others
1992	25.0	53.3	1.0	19.8	0.9
1993	30.7	48.1	3.4	17.0	0.8
1994	51.5	22.5	2.2	23.7	0.1
1995	54.5	13.9	7.8	23.1	0.7

Source: Taiwan ITRI, project ITIS. 1996.

Table 6.6 Major Indicators of IC Manufacturing Industry

Item \ Year	1991	1992	1993	1994	1995
No. of manufacturers*	10	10	10	11	12
Sales (US$ million)	622	870	1,566	2,642	4,505
Growth rate (%)	84.9	39.7	76.9	68.7	70.4
Advanced process capacity (μm)	0.8	0.8	0.6	0.4	0.3
Import/export	64:36	54:46	47:53	37:63	33:67
Investment/sales (%)	116.9	40.7	25.4	37.6	139.4
R&D/sales (%)	9.6	7.9	6.3	4.7	5.1
R&D/manpower (%)	13.3	11.2	7.2	10.3	9.3
Average R&D experience (year)	4.0	4.0	4.4	4.3	4.4

Note: Statistics do not include Vanguard Co.
Source: Taiwan ITRI, Project ITIS, 1996.

Table 6.7 Taiwan IC Manufacturers by Sales

Company	Sales (million $US dollar)		1994~1995 Growth rate (%)
	1994	1995	
TSMC	728	1,086	48.7
UMC	573	915	59.3
Winbond	324	690	114.4
TI-ACER	313	555	76.9
MOSEL-Vitelic	302	516	103.8
Micronix	211	335	61.9
Vanguard	-	196	-
Hualon	89	91	2.1
Holtek	69	91	31.7
EPISIL	19	25	34.6

Note: Exchange rate 26.42 (94), 26.48 (95)
Source: Taiwan ITRI, Project ITIS, 1996.

facilities. Winbond's expansion of the fab2 facility and the strong market demand for SRAM products made it one of the top three IC manufacturers in Taiwan. Mosel-Vitelic's new manufacturing plant, formally opened in July 1994, has maintained it within the top five IC manufacturers for two consecutive years. Except for TSMC, which is operated as a purely foundry service firm, memory products such as DRAM and SRAM are the major components of sales for all the rest of the top five manufacturers. If the current high demand for memory products continues, these companies will benefit in the near future.

In terms of overall product categories, Table 6.8 shows that memory products had the highest percentage and rate of growth between 1991 and 1995. Memory standard products account for more than two-third of Taiwan's total IC manufacturing output in 1994. Information-related products consist mostly of PC chipsets and computer peripheral ICs. The growth rate remains low in information ICs, the relative percentage of which dropped from 14.2 percent in 1993 to 8.0 percent of total IC manufacturing in 1995. Finally, only three domestic firms, TSMC, Holtek, and Hualon, are ASIC IC suppliers in Taiwan. The relative importance of this category dropped to 1.5 percent of overall IC manufacturing outputs in 1995.

The sources of Taiwan's IC manufacturing process foundry services are mostly provided by TSMC and UMC. Winbond and Hualon also provide little foundry services. Table 6.9 displays the sources of business for Taiwanese foundry services, which are mostly came from firms located in North America (55.5% in 1995). Taiwanese IC manufacturers are the second major source of demand for domestic foundry services. The relative percentage of domestic foundry services has continuously declined over the years from 1991's 47.4 percent to 1994's 30.5 percent, the lowest percentage since 1990. However, the trend has reversed in 1995. In 1994 Japanese firms increased their demand for Taiwanese firms' foundry services due to the appreciation of the Yen and the recognition of the quality of Taiwanese products and services, but the relative share declined in 1995 to only about 4 percent.

6.2.3. Wafer Manufacturing

Over the years, the continuous investment in IC manufacturing capacity by Taiwanese firms has never been disrupted. Domestic monthly production capability at the end of 1994, calculated in terms of 6-inch wafers, was approximately 278,000 pieces, a 42 percent increase over the previous year. Among Taiwanese IC manufacturers, most of them have the capability of producing 8-inch wafers, but the majority of the expanded output in 1994 and 1995 still came from the 6-inch wafer. The monthly production capacity and average productivity of wafer manufacturing in Taiwan since 1990 is summarized in Table 6.10. Among domestic wafer manufacturers, TSMC has the most impressive performance in output expansion. Winbond, UMC, and Macronix also have achieved significant growth over the years. Table 6.11 displays the current status of investment in 8-inch IC fabs by Taiwanese IC manufacturers.

Table 6.8 Taiwan IC Manufacturing Industry by Type of Business (Unit: %)

Item Year	Standard products				ASIC	Foundry	Other
	Information	Memory	Communication	Consumer			
91	15.9	32.7	7.0	26.8	9.6	8.0	-
92	15.4	40.2	6.3	19.2	5.3	11.0	2.6
93	14.2	51.2	8.5	10.9	5.0	10.2	-
94	10.7	59.4	7.3	9.1	2.2	10.1	1.2
95	8.0	67.1	2.4	5.7	1.5	14.3	1.0

Note: Statistics do not include sub-contractors
Source: Taiwan ITRI, Project ITIS, 1996.

Table 6.9 Sources and Relative Share of Demands for Taiwan's IC Manufacturing Foundry Services (Unit: %)

Year \ Area	Taiwan	Northern America	Western Europe	Others
1991	47.4	42.1	7.1	3.3
1992	48.0	45.0	7.0	0.0
1993	44.7	47.6	7.4	0.3
1994	30.5	55.1	5.1	9.3
1995	36.6	55.5	4.0	3.9

Source: Taiwan ITRI, Project ITIS, 1996.

During 1995, with the expansion of Vanguard's 8-inch fab, the completion of TSMC and UMC's fab3, and the operation of TI-Acer's fab2, the maturity of Taiwanese 8-inch wafer production finally arrived. It is estimated that the total output of 8-inch wafers by Taiwanese IC manufacturers could achieve 30 percent growth in 1996.

In 1995, Taiwan's IC industry still largely depends on the importation of wafers from foreign companies. Shin-Etsu of Japan, MEMC of the U.S., and Wacker of Europe are the three major long-term suppliers for Taiwan's IC manufacturers. Recently, Japanese wafer manufacturers, such as Sumitomo, Kamatsu, and Mitsubishi, have begun to pay more attention to the growth potential of Taiwanese manufacturing capabilities. Overall, Japanese firms have become the major foreign suppliers of Taiwan's IC industry, achieving 78.3 percent of the Taiwanese wafer market share in 1995 (66.2% in 1994). North America and Western Europe both equally accounted for 10.9 percent of wafer supply in 1995. As for the supply of MASK for IC manufacturing, domestic manufacturers accounted for more than 85 percent in 1995. The rest were almost all came from the United States.

TSMC is the largest foundry services company in the world, with 25 percent of the global share in 1996. Its total outputs of the 6-inch wafer was 1.2 million pieces in 1995, leading UMC by 40-50 thousands pieces a month. TSMC has two 6-inch and one 8-inch fabs in operation and several other 8-inch fabs under construction. In response to intensive competition, TSMC has accelerated its investment and significantly expanded production capacities in order to protect and extend its current industry leader status.

Table 6.10 IC Production Capacity and Average Productivity of Taiwan (Unit: 1,000 pieces)

Year	Item	Monthly production capacity	Average monthly production capacity
1991	4"	69	60
	5"	70	58
	6"	79	45
1992	4"	69	60
	5"	71	60
	6"	85	69
1993	4'	69	60
	5"	78	69
	6"	105	95
1994	4"	69	60
	5"	80	72
	6"	189	187
	8"	2	2
1995	4"	60	60
	5"	97	69
	6"	249	234
	8"	90	32

Source: Taiwan ITRI, Project ITIS, 1996.

Table 6.11 Investment of 8" IC Fabs in Taiwan

Company	Investment (US$ million)	Monthly capacity (thousand wafers)	Main products	Initial production
Vanguard	654.54	15	Memories	1994
TSMC	909.09	30	Foundry service	1995
UMC	909.09	30	CPU, ROM, SRAM	1995
TI-Acer	400	10	16M DRAM	1995
Nan-Ya	727.27	20	16M DRAM	1996
HMC	400	15	Memories, ASIC	1996
Winbond	1,818.18	60	Memories ASIC	1997
Macronix	1,090.90	40	Memories	1997
Mosel-Vitelic	909.09	25	Memories	1997
UMAX	909.09	25	Memories	1997

Based on US$1 = NT$27.5
Total amount of investment: 240 billion New Taiwan dollars
Total production capacity: 270 thousand wafers per month

Recent international cooperation in wafer manufacturing involve several Taiwanese semiconductor firms. First, UMC and Catalyst Semiconductor Inc. of Santa Clara, California, signed a foundry deal in 1996. Under this partnership, UMC will provide Catalyst with significant wafer foundry capacity for flash memory products, using a 0.5-micron process to be jointly developed by the two companies. UMC has also purchased a 10% equity interest in Catalyst to strengthen the relationship between the two companies.

Second, Union electronics Corp. (UEC) and International Micro Industries (IMI) of Mount Laurel, New Jersey, have formed a semiconductor wafer-bumping manufacturing joint venture in Taiwan, with technological know-how and an Industrial Property Right license from IMI. IMI will provide a turnkey operation, including equipment, materials, process skills, training, and follow-up technical support. The UEC manufacturing operation will provide world-wide merchant services for gold and solder bumps on 100-, 125-, 150-, and 200-mm wafers [Solid State Technology, 1996].

Finally, TSMC announced in June 1996 that it will form a US$ 1.2 billion equity joint venture agreement with three American firms: Analog Device (18% equity), Altera (18%), and Integrated Silicon Solution (4%), to establish a wafer foundry factory (WaferTech) in Camas, Washington. Overall, TSMC will contribute 57 percent of the equity share in this international joint venture. According to the operation plan, WaferTech will be able to produce ten thousand pieces of 8-inch wafer a month in the second quarter of 1998. And at the end of 1999, the projected output capacity will reach thirty thousand pieces a month. Initially, WaferTech's process technology will focus on $0.3\mu m$ and then toward the development of $0.25\mu m$ and $0.18\mu m$ technologies in the near future.

6.2.4. Microelectronic Packaging and Assembly

Taiwan's electronics industry started with foundry packaging services more than thirty years ago. Table 6.12 shows the progress of the Taiwanese microelectronic packaging industry over seven years. As of 1994, the total revenue of Taiwan's microelectronic packaging industry passed US$1.9 billion -- an annual growth rate of 36 percent. A majority of the packaging service (82%) is still as OEMs for companies outside Taiwan. However, since Taiwan's semiconductor industry is growing rapidly, more packaging services are geared toward local semiconductor manufacturers. In 1994, more than US$ 9.08 billion was invested on sub-micron 8" semiconductor fabrication by almost all of Taiwan's semiconductor manufacturers (see Table 6.11). It is anticipated that the Taiwanese semiconductor industry will have about 5% of the worldwide market share by the year 2000 and that the Taiwanese microelectronic packaging industry will grow rapidly along with the domestic semiconductor industry. Figure 6.7 illustrates the infrastructure of Taiwan's microelectronic packaging industry. The national strategy is to establish all aspects of these technical capabilities in order to be a world leader in the microelectronic packaging business.

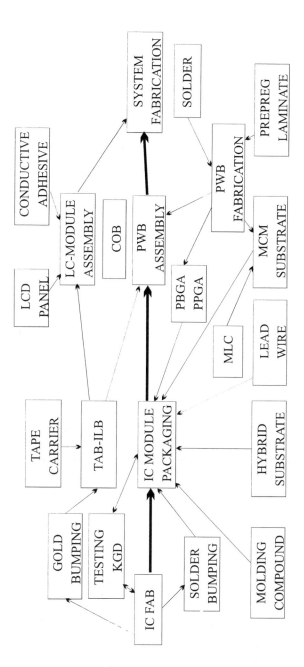

Figure 6.7 Taiwan's Microelectronic Packaging Industry Infrastructure

Table 6.12 Progress of Taiwan Microelectronic Packaging Industry

Year	1988	1989	1990	1991	1992	1993	1994
Companies	18	18	19	20	20	18	18
Production revenue (US$ million)	861	977	1,008	1,000	1,108	1,401	1,905
Annual growth rate (%)	27.6	13.5	4.2	-1.8	10.8	26.4	36.0
Export (%)	93	93	93	93	91	88	82

Source: Taiwan ITRI, Project ITIS, 1995.

The analysis of IC packaging activities in Taiwan can be categorized into foreign- and domestic-financed factories. Domestic-financed factories have aggressively expanded their capacities over the years. In 1995, there were twelve domestic-financed firms with total IC packaging sales of US$ 838 million, a 59 percent increase over 1994. Domestic wafer manufacturers have also managed to integrate with other IC testing and system manufacturing firms due to the increasing emphasis on turn key packaging services within the industry. On the other hand, several foreign-financed IC packaging firms in Taiwan have also expanded their capacities, and made improvements in management and operational strategy since 1994. In addition, they are now much more aggressive in joining domestic IC production and marketing activities.

Table 6.13 presents the percentage of IC packaging sales by Taiwanese domestic-financed firms in different regions. Taiwan area continues to be the largest sources of IC packaging revenues. It accounted for 59.7 percent of sales in 1995, followed by North America's 31 percent. It is expected that the share of domestic sales will increase as the result of the completions of many Taiwanese 8-inch wafer fabs in the next several years.

Table 6.14 shows the percentage of various types of IC packaging sales by Taiwanese domestic-financed firms for the period of 1991-1995. Among them, QFP has the highest rate of growth in relative share since 1991. In addition, it also has the largest percentage of sales revenues among all IC packaging services in 1995. This situation reflects the increasing demand by the downstream electronics industry for products with light, thin, short, and small design features.

Microelectronic packaging companies in Taiwan focused on the full rank of packages (see Table 6.15). To meet the requirements of advanced high-pin-count and high-speed ICs, the industry may follow its Japanese counterpart's road map and work toward fine-pitch packages such as plastic quad flat pack with a lead pitch of 0.3 mm or less.

6.3. Microprocessors

In 1992, the Taiwanese government began to promote the acquisition, transfer, and development of microcomputer CPU technology through private cooperations among electronics firms and research institutes. Since then, many microprocessor technologies, such as Sparc, MIPS, PA-RISC (Precision Architecture), and x86, have taken root in Taiwan.

Major representatives of Taiwan's microprocessor manufacturing companies are UMC, Winbond Electronics Corp., and Macronix International Co. Ltd. UMC acquired the SPARC microprocessor, which was developed by ITRI, and obtained 486 SX compatible microprocessors from Meridian (an American firm); Macronix and MIPS signed a contract to acquire MIPS R3000; and Winbond Electronics and Hewlett-Packard jointly introduced HP PA-RISC microprocessors.

Table 6.13 Percent and Region of IC Packaging Sales by Taiwanese Domestic-financed Firms (Unit: %)

Area / Year	Taiwan	Hong Kong Southeast Asia	Japan	Northern America	Europe
1992	53.5	0.2	1.5	32.1	12.7
1993	61.3	0.4	0.7	28.3	9.3
1994	59.5	0.2	1.3	32.1	7.0
1995	59.7	0.9	2.7	31.0	5.7

Source: Taiwan ITRI, ITIS project, 1996.

Table 6.14 Percent and Type of IC Packaging Sales by Taiwanese Domestic-financed Firms (Unit: %)

Type / Year	Plastic Packaging							Porcelain DIP	Other
	PTH		SMT						
	DIP	SK-DIP	SOP	QFP	PLCC	BGA	LOC		
1991	54.8	2.1	3.4	8.6	21.2	0	0	3.9	6.0
1992	41.1	6.2	9.7	23.8	14.9	0	0	2.3	0.5
1993	34.3	8.8	12.0	30.0	10.2	0	0	1.3	3.38
1994	25.4	10.4	15.8	36.5	9.2	0.4	1.1	0.9	0.3
1995	15.4	6.8	22.5	44.4	6.4	0.4	1.8	0.4	1.8

Source: Taiwan ITRI, ITIS project, 1995.

Table 6.15 Capability of Taiwan Microelectronic Packaging Industry (1993)

Package type	Maximum pin number	Companies
DIP	64	CET, OSE, SPIL
SOP	48	OSE
PLCC	84	SPIL, OSE, Lingshen, Caesar, Chantek, Talnet, Philip, CET
PQFP	304	ASE
PGA	309	TI
C-Quad	196	TI

In 1994, microprocessor products introduced by these three companies include: Macronix International's R#000/33 MHz (compatible with MIPS R3000), Windbond's W89K/66 MHz (compatible with Intel's 486DX), and UMC's U5S (compatible with 486SX). In addition to the development of microcomputer CPUs, in 1995, Macronix introduced a R3520 embedded processor, which is based on R3000 framework and peripheral logic, to be used in laser printer, digital still cameras, and other PDAs. Windbond also offered several new products based on PA-RISC technology. Finally, UMC has planned to establish R&D facilities in North America to develop next generation microprocessor.

In general, more advanced microprocessors, which target the markets of workstation, supercomputer, and high-end personal computers, are still dominantly produced by U.S. and Japanese firms. Taiwan's specialties are in the areas of personal computer microprocessors and other plug-in style application processors. Taiwan's microprocessor technology still lags behind advanced countries by one to two product generations.

Manufacturing processes, computer-aided design (CAD) tools, and microprocessor design capability are the major areas that require further research and development efforts by Taiwanese firms to catch up with American and Japanese companies. Through the use of strategic alliances and international joint ventures, Taiwanese firms have been able to significantly improve their capabilities in these areas.

6.4. PC Chipsets

The market for PC chipsets is very competitive. In just the last three years, three generations of chipsets have already become obsolete. Currently, the product life-cycle of PC chipsets is less than one year. Therefore, the ability to introduce new products in a short period of time ahead of industrial rivals has become vitally important. After facing severe competition in the world market for the last several years, Taiwan's products have finally gained a strong foothold in the global market.

In 1992, C&T Co. was the only PC chipset supplier in Taiwan. Since then, especially during the last two years, more than ten Taiwanese manufacturers have become chipset suppliers. Most of the current chipset manufacturers, such as Opti, ETEQ, Elite, and Suntac, are spin-offs firms from the C&T Co. Table 6.16 displays the operating status and market share of the major Taiwanese chipset companies in 1993. Before 1993, American companies such as C&T, VLSI, and Opti were the dominant players in the market. However, after three years of intensive efforts, Taiwanese manufacturers, such as UMC, ALI, SIS, and VIA, have finally taken over the leadership role in the market. In 1996, the markets for PC chipset in both Hong Kong and Taiwan are totally controlled by Taiwanese manufacturers.

Because the current type of personal computer adopts the chipset design concept and desktop PCs manufactured by Taiwanese companies have used very few foreign-made chipsets, it is not difficult to estimate the total market demand for Taiwanese chipsets by calculating the total output of desktop computers and PC motherboards. Tables 6.17 and 6.18 present the output status of Taiwan's desktops and motherboards in 1994, respectively. In addition, Table 6.19 displays Taiwan's major PC chipset vendors by type of PC motherboards. The production value of Taiwan's desktop computers (mainly 486/586-based CPUs) has reached US$ 4.325 billion, a 29 percent increase over the previous year, and 37 percent of the total information hardware production value. In addition, the output of portable computers has exceeded 2 million units in 1994, about 28 percent of the global market share. At the end of 1995, it was expected to surpass 32 percent of the worldwide market.

Taiwan's total output of PC motherboards in 1994 exceeded 80 percent of the global market share. In this area, Taiwan has ranked number one in the world since the 1980s. Among the various product lines in PC motherboards, the total output of the 486 CPU class already exceeds ten million units; the 586 CPU class reached one million units in 1994. Total output, including those manufactured in overseas Taiwanese facilities (90% of the Taiwanese overseas production is in mainland China), has reached 17 million units worldwide.

Many factors, such as efficient design and manufacturing capabilities, economies of scale in production, strong financial support from capital markets, and excellent customer service after sales, have contributed to the solid foundation for the personal computer and PC motherboard industry in Taiwan. Over the years, this strong industrial basis in Taiwan has attracted large amounts of ODM contracts from many world-leading PC manufacturers, including Compaq, Apple, and NEC.

In 1996, Taiwan's PC chipset suppliers, such as UMC, SIS, VIA, and ALI, ranked among the top seven chipset companies in the world. In addition, the total output by Taiwanese firms reached almost half of the global market share. Among domestic PC chipset companies, only UMC had a foundry factory; the rest were only design houses. Their competitiveness depends upon their design capability and time-to-market speed. All four companies are undertaking a two-tier strategy, simultaneously developing desktop and portable PC chipsets.

Table 6.16 1993 Operating Status of Taiwan's Chipsets Manufacturing Company

Ranking	Company	Product value (million $US)	Market share rate (%)	Quantity sold (thousand)	Market share rate (%)
1	VLSI	145	27	7,400	22
2	Opti	86	16	6,350	19
3	Acer	50	9	3,900	11
4	SIS	36	7	2,400	7
5	ACC	31	6	2,350	7
6	UMC	30	6	2,100	6
7	VIA	27	5	1,900	6
8	Symphony	24	5	1,800	5
9	IBM	98	19	1,200	4
10	C&T	24	5	1,140	3
11	ETEQ	18	3	1,100	3
12	Headland, LSI	20	4	900	3
13	Intel	65	12	850	3
14	Micro Silicon	17	3	600	2
	Others	87	16	3,350	10

Source: Dataquest, 1995.

Table 6.17 Statistics of 1994 Taiwan Personal Computer Produced Quantity (Unit: thousand)

CPU Type	1993		1994	
	Desktop	Portable	Desktop	Portable
386SX/SL	206	142	83	12
386DX	161	103	37	4
486SX/SL	1,023	581	1,121	965
486DX/DX2/ DX4	872	451	1,471	870
Pentium/586	22	14	340	25
PowerPC	-	-	27	181
Total	2,293	1,291	3,090	2,057

Source: Institute for Information Industry, 1995.

Table 6.18 1994 Taiwan Motherboard Production Status (Unit: thousand pieces)

Motherboard type	1991	1992	1993	1994
80286	1,321.1	226.6	19.7	20.1
30386SX	1,378.4	1,464.0	658.5	33.3
30386DX	1,286.3	1,752.4	2,695.7	395.5
80486SX	380.0	1,142.7	1,047.1	1,027.6
80486DX/DX2/DX4		1,376.3	4,712.4	9,009.3
Pentium	-	-	-	1,007.7
Other (Power PC MIPS)	-	-	-	35.0
Total	4,365.9	5,962.0	9,133.0	11,529.0

Source: Taiwan Institute for Information Industry, 1995.

Table 6.19 Taiwan 486/586 Chipsets Major Vendors

Vender	ISA+VESA	ISA+PCI	ISA+VESA +PCI	Notebook
SIS	486	486/586		486
UMC	486	486/586	586	486
VIA	486	586		
ALI	486	486/586	586	
Opti	486		486/586	486/586
Intel		486/586		
ACC				486/586
PicoPower				486
Green		586		
Unichip			486	
FOREX				586

Source: Taiwan ITRI, 1995.

6.5. DRAM/SRAM

Almost every major 16M DRAM supplier, including NEC, IBM-Siemens, Toshiba, Hitachi, Mitsubishi, and Texas Instruments, has developed 64M DRAM samples. The development of 64M DRAM technology can help to improve the existing manufacturing process for 16M DRAM products. At present, NEC is the world leader in 64M DRAM technology development. In the 256M DRAM area, Japanese firms are the most aggressive players, both in product development and manufacturing. It is projected that the first prototype of 256M DRAM could be introduced in 1996, and full commercial production could begin in 1997. Meanwhile, in Taiwan, successful experiences in the development of 4M and 16M DRAM by TSMC and UMC have encouraged many semiconductor firms, such as Vanguard, Nan-Ya, Powerchip, Mosel-Vitelic, and TI-Acer, to expand their investments in DRAM production.

Winbond Electronics will produce 16M DRAMs under license from Toshiba Corp. of Japan. Winbond expects to build a 30,000 wafer/month fabrication plant outside of Taipei by April 1997, with an investment of about NT$ 30 billion (US$ 1.2 billion). 64M DRAMs are also planned.

The introduction and subsequent high demand for Intel Pentium-based computers have accelerated the development of next-generation DRAM. In 1994, Mosel-Vitelic was the only domestic supplier of high-speed DRAM, accounting for only less than 6 percent of Taiwan's total demand. Table 6.20 shows the current status of DRAM manufacturing in Taiwan. It is vital for Taiwanese semiconductor manufacturers to develop 16M and 64M DRAMs in order to reduce the current heavy dependence on foreign supply.

Taiwan's total production value of DRAMs in 1994 was US$ 600 million, about 2.5 percent of global market share. The increase of production by Mosel-Vitelic, coupled with newly operated 8-inch wafer fabs invested by TI-Acer and Vanguard, has doubled the total production value in 1995 to US$ 1.25 billion. In 1996, UMC has begun to produce DRAMs. In addition, Powerchip and Nan-Ya's 8-inch wafer fabs are also scheduled to operate soon. As a result, Taiwan's total production value of DRAMs is expected to reach US$ 1.8 billion or 4.5 percent of the worldwide market.

Recent international technological cooperations in DRAM manufacturing include Mosel-Vitelic and the semiconductor group of Siemens AG of Germany. Both parties have signed a contract in the third quarter of 1996 to form a joint venture name ProMOS Technologies, which will be located at the HSIP to manufacture 64M and 256M DRAMs. Siemens will hold a 38 percent equity share, Mosel-Vitelic 62 percent. The first phase of the deal is worth about US$ 1 billion, with a subsequent stage adding another US$ 700 million. Both partners will contribute management and engineering resources towards operation of a fab currently being built by Mosel-Vitelic; it will initially work at 0.35 micron and later migrate to 0.25. The two firms will also set up a joint R&D center for DRAMs. Siemens will bring in its technological know-how in the field of memory components. Plans call for the new fab to employ about 1,500 employees after completion in 1998.

Because of the popularity of the Pentium (P54C) 3.3V, the demand for 3.3V-cache RAM is increasing in the market. In the past, Taiwan's semiconductor companies have successfully developed 3.3V 32Kx8 and 64Kx8 SRAM, the 15ns speed of which can satisfy the demands from downstream PC motherboard manufacturers. Currently, 0.5 μm is the most advanced SRAM product technology in Taiwan. Many Taiwanese manufacturers have attempted to develop 1M 32Kx32 synchronous-burst SRAM in order to improve CPU performance.

Taiwan's major SRAM manufacturers and their future R&D directions are presented in Table 6.21. In the high-speed SRAM product market, Taiwanese manufacturers account for more than 65 percent of total domestic demands. Over the years, Taiwan's SRAM manufacturers have achieved many significant improvements in design capacity, product development, quality improvement, and manufacturing processes. Taiwan's high-speed SRAM products, especially in PC applications, have also achieved high-standard global status. Many leading desktop PC/motherboard makers, such as Acer, FIC, Intel, IBM, Packard Bell, Compaq, and AT&T, have used Taiwanese cache memory in their products. In 1995, due to the strong demand in cache RAM, the total production value of Taiwanese made SRAM reached US$ 869 million, a 160 percent increase over

1994. However, the severe price competition will slow the rate of growth in 1996. Nevertheless, it is forecast that the global market share of Taiwan's high-speed SRAM products will continue to grow in the years to come.

6.6. Non-volatile Memory

Memory can be categorized into volatile and non-volatile (NV) memory, according to the way it accesses and stores data. Volatile memory, which can be further classified into DRAM and SRAM, must constantly be fed by a power supply in order to store data. On the other hand, non-volatile memory can store data even when electrical power is turned off. NV memory can be classified as ROM, EPROM, EEPROM, and Flash, according to whether the stored data can be erased and the way it is erased. Table 6.22 presents the global status of Taiwan's memory-chip industry for the years of 1991, 1994, and 1995. The production of volatile memory, especially SRAM, has grown tremendously over the past several years.

Currently, the only domestic manufacturers of Mask ROM in Taiwan are Macronix and UMC. Both of them have already successfully manufactured 32M products, and the prototype of a 64M product was introduced in 1995. In 1994, the top four manufacturers in the world were Sharp, NEC, Toshiba, and Samsung. Taiwan's Macronix was ranked sixth in the global market.

Macronix is the only EPROM manufacturer in Taiwan. Its global market share was ranked seventh in 1994. Currently, in the global EPROM market, SGS-Thomson and Texas Instruments are the two major competitors of Macronix. Because of the continuous appreciation of the Japanese Yen and the increased cost of input for production, the estimated price of EPROM in 1995 could increase about 15 to 20 percent over the current price . As for EEPROM, with the exception of ISSI, which provides a limited supply to the domestic market, almost all EEPROM memory supplies are coming from U.S. manufacturers such as Amtel, Catalyst, Microchip, NS, and Xicor.

Macronix is also the primary domestic supplier of Flash in Taiwan. Its products include 1M, 4M, and 16M. Foreign suppliers of Flash to Taiwan include Intel, AMD, Catalyst, and SGS. Macronix planned to introduce 2M and 8M products in 1995. In addition, Winbond is projected to introduce 256K Flash into the market in the near future; and TSMC and UMC will manufacture Flash memory through foundry service contracts. ISSI is still the primary manufacturer of 1M Flash. The manufacturing of 4M and 8M Flash are also in ISSI's future plans. Overall, it seems Macronix will be able to maintain its leadership status in the manufacturing of Flash for the domestic market.

Table 6.23 compares the status of Taiwan's NV memory technology with current advanced technology. It shows that Taiwan's Mask ROM and EPROM technologies have already achieved parity with advanced countries. Although Taiwan's Flash memory technology still lags behind the current advanced technology, it is very likely that Taiwan would be able to catch up with advanced countries in a couple of years.

Table 6.20 Status of DRAM Manufacturing in Taiwan

Company	Factory	Wafer size (inch)	Production capacity (µm)	Capacity (kilo pieces)	Major product	Manufacturing status
Mosel-Vitelic	Fab1 Fab2	6" 8"	0.55 0.35	15 25	1M/2M/4M 16M	In manufacturing. Will be 30K/month in 1996. 5K/month experimental productivity in first half of 1997.
TI-Acer	Fab1A Fab1B Fab2	6" 8" 8"	0.5 0.45–0.35 0.35	18 15 25	4M 16M 16M/64M	In manufacturing. July 1996 starts. At the end of the year will be 8K. Built at 1995. Q2 will be manufactured in 1997.
Vanguard	Fab1A Fab1B	8" 8"	0.5 0.4	14 15	4M 16M	In manufacturing. Will be 14K/month at the end of the year. Built in July, 1995. Will start manufacturing in 1997.
Powerchip	Fab1	8"	0.4	25	16M	1996, Q3 will start. Will be 15K/month in 1997.
Nan-Ya	Fab1	8"	0.45	18	16M/64M	Will start at the end of 1996. Monthly production will be 5K.

Source: Taiwan ITRI, ITIS project, 1995.

Table 6.21 Taiwan's Major SRAM Manufacturing Products and R&D Direction

Company	1994 New products	Future R&D
UMC	64K×16 Synch SRAM 0.5µm 32K×8 5V SRAM 32K×8 3.5V SRAM 32K×8 5V SRAM	32K×32 Synch burst SRAM 64K×8 3.3V SRAM 0.45µm SRAM
Winbond	0.5µm 32K×8/64×8 - 15ns 5V/3.3V/mixed SRAM	32K×32 Synch. burst SRAM
Etron	0.5µm 32K×8 - 12/15ns 5V/mixed SRAM	32K×32 Synch. burst SRAM
UTRON	0.5µm 32K×8 - 70/100ns SRAM	1M 128×8 Low Speed SRAM

Source: Taiwan ITRI, ITIS Project, 1995.

Table 6.22 Global Market Share of Taiwan's Memory Chip Industry (Unit: %)

Product	Year	1991	1994	1995 (estimated)
DRAM	Market	6.1	7.4	7.4
	Product value	1.4	2.5	2.8
SRAM	Market	10.2	8.2	13.9
	Product value	4.5	7.4	16.3
ROM	Market	6.8	5.9	-
	Product value	9.4	8.7	7.9
EPROM	Market	3.7	5.6	-
	Product value	0	5.0	7

Source:Dataquest; Taiwan ITRI, ITIS project, 1995.

Table 6.23 Comparisons of Non-Volatile Memory Technology

Product	Taiwan's current technology		Advanced technology	
	Capacity	Manufacturing process	Capacity	Manufacturing process
ROM	32M in production 64M sample	0.45 Fm	32M in production 64M sample	0.45 Fm
EPROM	8M	0.60 Fm	8M	0.50 Fm
EEPROM	4K	-	1M	-
Flash	4M in production 16M sample	0.60 Fm	16M in production 32M sample	0.50 Fm

Source: 1995 Taiwan Semiconductor Year Book, Page 6-55

Table 6.24 Taiwan's Non-Volatile Memory Total Sales (Unit: %)

	Sales (US$ million)	Percentage (%)
NV Memory	253.0	100.0
(1). ROM	130.0	51.4
(2). EPROM	86.5	34.2
(3). EEPROM	17.5	6.9
(4). Flash	19.0	7.5

Source: Taiwan ITRI, ITIS project, 1995.

Table 6.24 presents Taiwan's total production value of NV memory in 1994. The overall value could reach an estimated US$ 253 million -- about 5 percent of the global market share. Among the statistics, ROM accounts for the largest portion of total value (51.4%), followed by EPROM memory (34.2%). Finally, Figures 6.8 and 6.9 illustrate the market and production values of Taiwan's memory chip market. In 1995, NV memory products only accounted for 6 percent and 14.4 percent of the market and production values, respectively, down from 1994's share of 10.8 percent (US$ 253 million) and 22.4 percent (US$ 265 million). SRAM shares were increased in both values almost exactly at the expense of NV Memory. DRAM still retains the largest market share (75.2%) and highest percentage of production value (50.5%) in Taiwan's memory chip market.

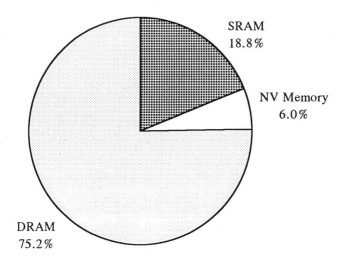

Total market value: US$ 4,082 million

Figure 6.8 1995 Taiwan Memory Chip Market

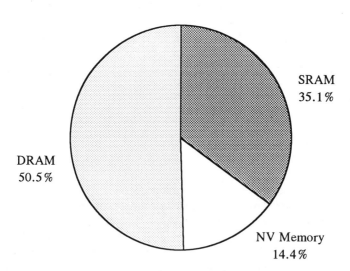

Total production value: US$ 2,470 million

Figure 6.9 1995 Taiwan Memory Chip Production Value

6.7. ASIC

ASIC products have become increasingly popular in Taiwan. As a result, competition has intensified among foreign firms who are trying to exploit the Taiwanese market. Major industrial applications are mostly in the data processing area (it accounts for 70%), such as PC chipsets, PC I/O chip, graphics chip, and network adapter. Other specific applications also receiving significant attention are in the areas of communication and consumer electronics, such as fax, pager, language translation machine, and personal digital assistants (PDA).

Taiwan's ASIC product supplies are mostly coming from Japanese firms, such as Epson, Toshiba, and NEC. Korean firms, such as Samsung, also provides limited supplies to Taiwanese firms. As for types of ASIC products, the majority of ASIC suppliers are concentrated on gate array products. This is different from the practice of American companies. Traditionally, American ASIC firms pay more attention to the standard cell products. However, as mentioned above, Taiwanese firms are mostly SMEs facing serious financial constraints and are more concerned about costs and product delivery schedules. Therefore, when they are choosing foreign partners for the development of ASIC products, Taiwanese companies almost always give first consideration to Japanese manufacturers.

6.8. Discrete Components

Taiwan's discrete component industry can be divided into three categories: transistors, diodes, and thyratrons.

6.8.1. Transistors

The market for transistors has been growing significantly due to increasing demands for appliances since the global economy began to recover in the early 1990s. According to the statistics estimated by the World Semiconductor Trade Statistics (WSTS), the global sales volume of transistors in 1994 exceeded US$ 6 billion, a 20 percent increase over 1993. Furthermore, the market growth rate in 1994 was the highest in the last three years.

In 1994, Taiwan was able to generate US$ 250 million in sales volume and produce 2.3 billion transistors, with the total number of manufacturers unchanged and production capabilities expanded. The growth rates of Taiwan's sales volume and production volume in 1994 were 36% and 24%, respectively. Compared with global transistor industry development, Taiwan has shown a very impressive performance.

In general, Taiwanese manufacturers, such as High-Sincerity Microelectronics Corp., only supply a small portion of transistors to the domestic market. Foreign manufacturers are still the major suppliers for the domestic demand for transistors. In terms of sales volume, Toshiba and International Rectifier (IR) were the leading suppliers for Taiwan's transistor market in 1994, followed by Sanyo, Hitachi, and Philips. As a group, Motorola and NEC were a distant third. As for product segmentation, IR, Toshiba, and Fuji are the dominant players in the area of MOSFET power transistors, while Sanyo, Hitachi, and Philips lead in producing bipolar power transistors. Finally, NEC, Motorola, Philips, and Toshiba specialize in small-signal transistors.

In addition to Hi-Sincerity and MOSPEC, the two major Taiwanese manufacturers of transistors, Hitachi, Siliconix, and Sanyo also have established transistor packaging factories in the export-processing zones of Taichung and Kaohsiung. According to data from the Statistics Bureau of the Ministry of Economic Affairs, Taiwan's annual production volume of transistors has been increasing since 1990, with more than 2.3 billion units manufactured in 1994. The average annual rate of growth between 1990 and 1994 was 19.3 percent. The annual domestic sales volume of transistors also increased to US$ 253.6 million in 1994. The average annual rate of growth between 1990 and 1994 was 24.2 percent.

According to data from Taiwan's Customs Bureau, three foreign-owned packaging manufacturers have been ranked within the top three in product exports for a long time. Their combined total exports account for 95 percent of Taiwan's total export value in transistors. This implies that domestic transistor manufacturers still have significant opportunities to grow.

Overall, only two domestic companies are devoted to transistor product research and development. It is obvious that Taiwan's transistor product lines and technology are still far behind the advanced countries. Taiwan must integrate necessary resources in order to develop advanced transistor technologies and products, and become more competitive.

6.8.2. Diodes

The diode industry is crucial in the overall development of Taiwan's semiconductor industry. Even though it no longer represents the mainstream of Taiwan's semiconductor industry, during its peak in the 1970s and 1980s, the diode industry contributed significantly to generating tremendous foreign reserves, creating job opportunities, and providing training for technicians. In addition, the diode industry stimulated the development of upstream and downstream industries.

Rectifying diodes can be applied to a wide range of products, including missiles and automobiles. Their major application is in electrical appliances such as televisions, computers, and power supplies, and in other electrical equipment. The diode is well-known as an active electronic component in converting alternating current to direct current.

Taiwan's rectifying diode industry started in 1969, with the establishment of the General Instrument (GI) factory (a U.S.-owned facility). In 1977, Rectron Ltd., a domestic-owned company, started its production in the Tu-Chen industrial zone located in Taipei Hsien. Twenty years later, Taiwan has more than fifteen rectifying diode factories, with stable positions in peripheral equipment, material, and human resources. The production of rectifying diodes has been flourishing, and they have become a major product of Taiwan's electronic industry. As for the production volume of diodes, Taiwan's GI, other global leaders in diode production, technology, and marketing, and Taiwanese manufacturers (including factories located in China and Malaysia), account for about 60 percent of the global production volume in generic rectifying diodes.

For the past few years, Taiwan's diode industry suffered from the appreciation of New Taiwan Dollars, the increase of labor costs, the shortage of manpower, and the failure to introduce new products. All these factors have reduced Taiwan's diode industry's profit margin and global competitive advantage. There has been no new factory established in Taiwan for the past two years. Meanwhile, the leading manufacturers in the industry, such as GI, Rectron, and Goldentech Discrete Semiconductor, Inc., have built their new factories in Malaysia. However, since the industrial environment in Malaysia is still inadequate, both Rectron and Goldentech have suffered in substantial losses. In response, almost every one of Taiwan's diode manufacturers, such as Rectron, Taiwan Semiconductor Co. Ltd., Goodark Electronic Corp., and Allied Electronic and Semiconductor Technology Inc., has transferred part of its production facilities to the Chinese mainland to pursue low-cost labor. However, there are too many players in the diode market; as a result, the profit margins for diode manufacturers are very thin. If this condition remains unchanged in the near future, it could lead to a financial crisis for many firms. Consequently, corporate mergers and alliances are expected among existing diode manufacturers.

The impacts from the appreciation of NT dollars and the continuous increase of labor costs will surely lead to a more hostile environment for diode exports. Domestic manufacturers must reevaluate their business strategies and adjust their corporate structures in order to sustain a competitive advantage. By speeding up the process of transferring factories to low-cost areas in Asia, enhancing production automation, altering corporate structure, and improving product and technology development, Taiwan's diode manufacturers should be able to cut their production costs and improve their efficiencies. In general, Taiwan's diode industry should look into the following issues to ensure future prosperity in the diode business:

- Several leading global semiconductor manufacturers, such as North America's Motorola, Europe's Philips, and Japan's Toshiba, tend to place their generic rectifying diode orders with OEM manufacturers to cut down their manufacturing costs. This could create contract opportunities for Taiwan's diode manufacturers. It could also help to maintain the current capabilities and enhance future development possibilities of the domestic diode industry.

- The upcoming threats from South East Asia and mainland China's diode industries, which were the necessary consequences from the outward movement of Taiwan's semiconductor industry, have created turbulence in the already highly competitive diode business. Among these emerging competitors, the Chinese mainland has strong advantages in having low labor costs and a large domestic market. Even though its ability to compete in the global market is still quite fragile, mainland China should become a major threat to Taiwan's diode industry in the near future. Therefore, Taiwan must develop a long-term industrial strategy to counter the threat.

• Most well-established global manufacturers have been gradually transforming their product planning strategy and production patterns. The trend is to abandon the production of low-value-added, low-profit-margin products, while still trying to hold on to market share. The leading global manufacturers have aggressively developed high-value-added products, such as the high-speed restore diode for the information industry, the high-power diode, and a diode for the automobile industry. In the short term, Taiwan's diode industry will not be directly affected by this transformation. However, in order to accelerate the process of upgrading the domestic industry and to ensure future development, Taiwan's diode industries must speed up the development of new markets, applications, and product manufacturing technologies.

6.8.3. Thyratron

The development of IC products has always been the mainstream of Taiwan's semiconductor industry. The development of thyratron and other discrete component technologies has not experienced such good fortune. Except for the rectifying diode, other discrete component industries have all had difficulties meeting expectations.

Currently, Photron Semiconductor Corp. is the only domestic supplier of thyratron products. The rest of the supply is coming from foreign manufacturers. Photron's product lines include SCR, TRIAC, and DIAC. Major foreign suppliers of thyratron products include U.S.-based Motorola and Teccor, Japan-based Toshiba and Mitsubishi, and Europe-based SGS-Thompson and TAG. According to WSTS's analysis, the sales volume of the global thyratron market exceeded US$ 700 million in 1993, about 20 percent more than the previous year.

Taiwan's domestic demand for imported thyratron products was about US$ 22.6 million in 1994, a 5.9 percent increase over 1993. Overall, Taiwan's demand for thyratron products accounts for 3 percent of the total global sales volume.

Thyratron is a relatively mature technology and its manufacturing process is fairly stable, so the existing manufacturers have already accumulated much experience. Under these circumstances, it is rather difficult for new players to enter this market. Photron is the only Taiwanese firm to successfully produce thyratron products. Taiwanese firms may have opportunities to produce thyratron products in the near future when well-established manufacturers in advanced countries decide to transfer their production facilities overseas.

6.9. Multi-chip Modules

Taiwan is involved in the development of multi-chip module technologies for various product applications. To promote a Taiwan MCM roadmap, ERSO set up an MCM Service Center in the fourth quarter of 1994 to provide total response to customers' needs in advanced packaging. The MCM Service Center can design MCMs based on customers' net lists, block diagrams, and bare chip information. After a review of the MCM's proposed electrical, thermal, and physical designs with the customer, the design is turned into a series of fabrication steps, including substrate processing, bare chip assembly, and module functional testing. At the customer's request, the MCMs also can be subject to a set of reliability tests to ensure their quality and long-term performance.

ERSO has formed a strategic alliance partnership with Intel (Patterson 1994) to provide known good bare chips, including the most advanced Pentium CPU chip, with the reliability and price parity of conventional packaged chips. In turn, ERSO will fabricate MCM prototypes for Intel and its customers. In addition to using bare chip suppliers worldwide [Hagge and Wager, 1995], ERSO plans to work with Taiwanese IC manufacturers on bare chip testing and burn-in technical issues, as well as to form a known good die(KGD) alliance. For future technological advancement and growth of professional packaging engineer supply, ERSO also plans to work with six universities, (National Taiwan University, National Chiao Tung University, National Tsing Hua University, National Cheng Kung University, National Central University and National Kaohsiung Institute of Technology), which are conducting MCM research programs sponsored by the National Science Council. Developments in Taiwan's MCM technology include:

- In April 1993, ERSO received an MCM-D technology from PMC-Sierra Inc [Costlow and Patterson, 1993]. A maximum of five-layer interconnect structure, which consists of aluminum conductor lines and polymide dielectric, is being built on a silicon wafer. The technology utilized wirebonding for bare chip assembly.

- In August 1993, ERSO started a joint program with four major packaging companies in Taiwan, (Philips, AES, Caesar and Siliconware) to develop low-cost PQFPs for MCM applications, based on Taiwanese capacity and experience in PQFP manufacturing, and because the package format is compatible with the surface-mount (SMT) assembly lines in Taiwan.

- Ball-grid array (BGA) and 3-dimensional packaging are under development by ERSO. In the near future, Taiwanese companies will adapt PWB technologies to further reduce MCM costs, as well as developing new testing methodologies, fixtures, and IC testers. As shown in Table 6.25, ERSO has also been developing other low-cost and high-performance technologies, such as flip-chip assembly and wafer substrates integrated with thin-film resistors and capacitors.

Table 6.25 MCM-D Technology Developments at ERSO

Technology development	Interconnect substrate	Bare chip assembly	Next level package
Phase I (1993)	5-layer Al/Polymide on Si wafer	Wire bonding	PQFP
Phase II (1994-1995)	Integrated the above substrate with thin film resistor and capacitor	Flip chip	BGA, 3-D package

Source: Taiwan ITRI ITIS project, 1995.

Chapter 7

FLAT PANEL DISPLAY

Flat panel displays (FPDs) are increasingly important in information electronics products. Compared with the CRT (cathode ray tube) used in traditional televisions, FPDs are thin, lightweight, and power efficient devices that present images without the bulk of a picture tube. As a result, FPDs represent a large and rapidly growing industry worldwide, and are expanding into an increasingly diverse set of systems. The largest demand for FPDs is for use in computers, mainly portable systems such as laptops, notebooks, and handheld PDAs. These electronics devices and many other consumer electronics use a type of FPD called liquid crystal displays (LCDs).

7.1. Industry Overview

The global FPD industry uses a diverse set of technologies to satisfy a broad array of applications. The dominant technology is the LCD, which itself comes in many forms; the primary variations are the more advanced active matrix LCD (AMLCD) and the basic passive matrix LCD (PMLCD). Measured by value of sales, LCDs account for approximately 87 percent of the worldwide FPD market in 1995 (US$ 10 billion), evenly divided between active and passive matrix types. The FPD market as a whole is projected to double between 1995 and 2001, and AMLCDs are expected to account for 54 percent of total FPD market [Office of Technology Assessment, 1995]. In addition to LCDs, smaller shares are accounted for by another two type of FPD: plasma displays and electroluminescent (EL) displays. In terms of value, these four FPD types make up the vast majority of the FPDs currently in use.

In addition to Japan, many east Asian countries are entering into the race to develop AMLCD production capability. Firms in South Korea, such as Samsung and Hyundai, are leading the race among Asian NICs. Korean and Taiwanese firms entered the FPD industry for different reasons. In general, Korean firms appear to view FPDs as an important industry on its own as a potential successor industry to CRTs (mainly for exporting purpose). In addition, FPDs are also viewed as a companion industry to semiconductors which could take advantage of the existing manufacturing infrastructure. On the other hand, the drive to develop FPD manufacturing capabilities in Taiwan appears to be related to its role as a main component in personal computer manufacturing. Taiwanese firms have the largest worldwide market share of computer monitors (see Table 5.2), and have a growing share of the portable computer market. An insufficient supply of LCD screens during 1994 meant that Taiwanese manufacturers were unable to fill many orders. Therefore, Taiwan appears determined to become more independent of FPDs supplied by Japan.

Because of the increasingly large demand for portable computer and multimedia products, many electronics firms, such as Sharp, NEC, Toshiba, Hitachi, and Fujitsu of Japan, and Samsung, Hyundai and Lucky Goldstar of South Korea, are devoted to the efforts of expanding their LCD manufacturing facilities. Most of these firms are invested in TFT LCDs (AMLCD form). As for Taiwan, mass production capabilities of large screen AMLCDs will not be ready until 1997. In 1996, there are several manufacturers of TN and STN LCDs (PMLCD form) in Taiwan, including Picvue, Nan-Ya Plastics, and Chunghwa Picture Tubes (a subsidiary of Tatung), whereas the production of small screen AMLCDs in Taiwan has been led by two firms: Unipac and PrimeView International.

7.2. LCD Technology Development in Taiwan

Although the decision to invest in high-end LCD products by Taiwanese firms lags several years behind that of Japan and even Korea, the potential for high rates of growth in Taiwan's LCD industry remains strong. The early LCD industry in Taiwan produced only low-price, low-technology TN LCDs. As the world-wide demand for displays increase rapidly, Taiwanese manufacturers began to invest in high-tech, high-value-added LCD products, such as STN and TFT LCDs. Table 7.1 summarizes the current status of Taiwan's display industry including both CRT and LCD products.

ITRI was the first Taiwanese research institute to conduct high-tech LCD research and development. It has more than eight years of R&D experience began from the 3-inch, 89,000 pixel TFT LCD to today's 345,000 pixel, 16:9 projection television. In 1994, ITRI experimented with 6-inch substrate to manufacture a small quantity of the 3-inch, 307,000 pixel TFT LCDs. This experiment has established ITRI's capability to provide a "base line" necessary for further R&D efforts. In 1995, ITRI used a $300 \times 300mm^2$ large glass to build the 10.4-inch TFT LCD base line. ITRI also developed a reworkable chip-on-glass (COG) process in 1994 to solve the problem of high-cost COG manufacturing. This technological breakthrough has greatly improved the domestic manufacturing technology, and has made COG process technology more practical in high-density LCD manufacturing.

Table 7.1 Taiwan's Display Technology

Category	Size	R & D	Pilot production	Mass production
CRT	14" CRT 15" FS 21" CRT Over 28" 16:9 WTV	X	X	X X X
LCD	STN 3" - 10" B/W STN 10" color TFT under 6" TFT 10" color TFT over 10"	X	X X	X X

Source: Taiwan ITRI, project ITIS, 1996.

Overall, Taiwan's ERSO has worked with companies to develop various prototype of FPDs, and has also been a source of trained engineers for companies such as PrimeView. In addition to the efforts of developing AMLCD products, ERSO's FPD project also includes efforts to develop other types of FPD technologies such as plasma and FED (field emission display).

As for the investment strategy of Taiwanese FPD firms in 1996, TN LCD and monochrome STN LCD are still the two major products capable of self-development. In the short term, many important parts, such as driver chips, color filters, and backlights, still need to be imported from foreign sources, especially from Japan. The lack of a materials and equipment infrastructure in Taiwan has kept the cost of production high because spending on equipment and components comprises the majority of FPD manufacturing costs.

Taiwanese firms are trying to develop an indigenous supply of notebook screens to lessen their dependence on foreign-made displays. In order to reduce the production cost and stabilize the manufacturing process, the development of LCD peripheral industries is required to maintain the global competitiveness of Taiwanese LCD firms. Recent efforts to achieve self-sufficiency already demonstrate some progress. For example, in the color filer area, Unipac is able to make color filters for its own TFT LCD products. UMC is conducting intensive R&D efforts on driver IC. Furthermore, Chungtex Electronic Co. has already begun to produce small quantities of backlights since 1995.

7.3. Major Taiwanese LCD Manufacturers

In 1996, three major domestic manufacturers (Picvue Electronics Ltd., Nan-Ya Plastics, and Chunghwa Picture Tubes, Ltd.) are developing color STN products. For TFT LCDs, although several manufacturers have planned to develop and produce commercial products, only PrimeView and Unipac were actually in production. The current product development status of each major Taiwanese LCD manufacturer is discussed below.

Picvue Electronics is the first Taiwanese company to produce STN LCD. Its initial strategy for LCD development was based on TN LCDs. In 1996, Picvue moved into the STN field and also planned to go further into the production of TFT LCDs. The source of Picvue's TFT LCD technology is mostly coming from its U.S. subsidiary -- the Polytronix Co. Since the development of TFT LCDs requires large capital investment, Picvue plans to establish a technical alliance among domestic and foreign firms to build a TFT LCD factory in Taiwan ready for operation within three to five years.

Overall, TN LCD products accounted for 85 percent of Picvue's total sales in 1994. In addition to the basic TN LCD products, Picvue applies a three π-cells combination to make light covers that can convert monochrome CRT into color CRT. Picvue also uses π-cell with reaction speeds below one μ second to produce surface-mode TN LCDs. In addition to its current language translator and electronic dictionary, Picvue will try to introduce a Quaiter-VGA (320×240, 240×64, 192×69) personal digital processor and a 9.4-inch to 10.4-inch class of monochrome STN LCD. As for color STN LCD, Picvue currently is implementing very intensive R&D efforts. The prototype is projected to be ready at early 1996.

Nan-Ya's Su-lin laboratory has made progress in research and development of small and medium sized STN LCDs. Nan-Ya's Gin-Hsin factory, which is located in Tao-yun county, has already begun the mass production of STN LCDs (5.6 inches) in March 1995. The factory is capable of producing 500,000 LCDs annually for notebook computers, and supplying another 2 million LCDs to the downstream factories for various purposes. As for the production of TFT LCDs, Nan-Ya is planning to cooperate with Japanese conglomerates, such as Mitsubishi and SEKIO-EPSON, to develop color filters and 10.4-inch TFT LCDs, rather than solely relying on its own R&D efforts.

Chunghwa Picture Tubes, which is the largest domestic producers of CRTs, is also aggressively striving toward the production of LCDs. The licensed technology which enables Chunghwa to manufacture large volume of both color and monochrome STN LCDs was acquired from Toshiba of Japan. It is expected that Chunghwa will soon be able to introduce the 10.4-inch STN LCDs to the market in the early 1997. In spite of progress made in STN LCDs, the decision of whether to go into the production of TFT LCDs is still under evaluation.

PrimeView was one of the two Taiwanese manufacturers who pioneered in producing TFT LCDs. In early years, PrimeView signed an agreement with ITRI to acquire TFT process technology and to obtain manpower training in LCD manufacturing. The US$ 160 million TFT LCD plant, which is located in the newly expanded section of the HSIP (3rd campus), completed construction at the end of 1995. So far, it is the largest domestic LCD factory and is expected to be fully operational in the third quarter of 1996 after several adjustments for mass production have been made. The initial monthly capacity is projected to be 6,000 pieces of 370mm×470mm size of glass substrate. After the completion of this project -- an estimate total investment of US$ 283 million -- an additional monthly capacity of manufacturing 18,000 pieces of 550mm×650mm size of glass substrate will be added to the total monthly outputs. It will make PrimeView the first domestic manufacturer to be able to supply large quantities of TFT LCD substrate.

Unipac Optoelectronics was the earliest Taiwanese manufacturer of TFT LCDs. It was a joint venture established in July 1990 among several major Taiwanese semiconductors firms, namely UMC, TECO Electric and Machinery Co., and Sampo Technology Corp. In early 1996, Unipac's fab1, which is leased from UMC, has a monthly capacity of manufacturing about 3,000 pieces of 370mm×470mm size of glass substrate for use in 4- to 6-inch TFT LCD products. Due to the space constraint in fab1, Unipac began to construct the fab2 in 1994 within the 3rd campus of HSIP. Upon completion, the monthly production capacity of fab2 is projected to be 40,000 pieces of 370mm×470mm size of glass substrate. Although Unipac already has the capability of manufacturing 9.5-inch TFT LCDs, the product lines in 1996 still mainly consist of 4- and 5-inch TFT LCDs.

7.4. Product Market Outlook

Because of intensive global competition and newly expanded production capacities of many FPD firms, both domestic and foreign, the prices of LCDs have been plummeting since 1995, especially the 10.4" TFT LCDs. This will have a large effect on the supply of STN LCDs and subsequently influence the price of color notebook computers.

Although the decision to invest in high-end LCD products by Taiwanese firms lags several years behind that of Japan and even Korea, the potential for high rates of growth in Taiwan's LCD industry remains strong. The reasons for optimism include:

- Domestic demand for LCDs will remain strong, especially from Taiwan's dominant personal computer industry.
- Taiwan's process technology for manufacturing semiconductors has improved to the point where it can assist similar research and development processes of the LCD industry and provide low-cost IC drivers for the LCD products.
- The investment timing is proper. Taiwan's late entry into the LCD market has significantly reduced the initial cost of development.
- TFT LCD has been selected by the Taiwanese government as one of the strategic technologies for future development. Figure 7.1 indicates that

Taiwan's LCD industry has gradually becoming an integrated system due to strong government and academia supports, as well as the formulation and implementation of industry competitive strategy.

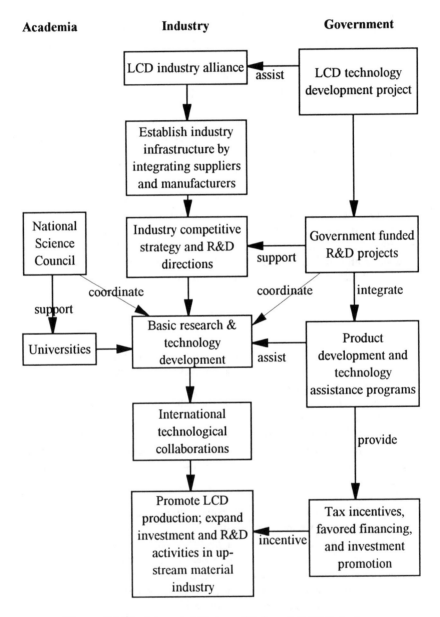

Figure 7.1 The integrated System of Taiwan's LCD Industry

Chapter 8

INDUSTRIAL STRUCTURE AND PERFORMANCE

8.1. Industry Structure

Taiwan's electronics industry is characterized by many small and medium enterprises (SMEs), especially in the downstream end product market where industry concentration is low. The most important structural factor determining industry performance in this area is the ability of a firm to enter the industry. The barriers to entry are relatively low. However, in the electronic components segment, barriers to entry can be very high in some areas such as wafer fabrication. The worldwide electronics (computer) industry was originally dominated by a small number of vertically integrated companies, such as IBM and Digital. Over the years, these companies faced fierce competition from new entrants into the field, and this change ultimately transformed the structure of the industry into a horizontally layered one with numerous players in each layer.

The relationship between the layers (e.g., buyer-supplier relationship) is relatively open in that firms can freely choose the supplier who often offers what they need at the best price. Taiwanese firms have taken advantage of this trend and developed multi-layer system of specialists, but Taiwan faces stiff competition in each layer from South Korea, Malaysia, and China.

8.2. Industry Financial Performance

A one-year accounting of rate of return for the major companies is presented here in order to give a general view of the performance of Taiwan's electronics industry in 1995. The major financial performance indicators of Taiwan's leading information electronics, communications, and electronics parts and components manufacturers are included in the Appendix.

Figure 8.1 illustrates the profitability of eighty-nine of Taiwan's top one hundred information and communications equipment manufacturers in 1995 (not including those that did not report either assets or after-tax profits). However,

Figure 8.1 does not include Kunnan Enterprise, Ltd. because it reported an extraordinary 73 percent drop in its rate of sales growth and a negative 90 percent rate of return on assets (ROA).

In 1995, the majority of Taiwan's information and communications equipment makers had annual sales under US$ 200 million. Except for a dozen manufacturers, most of the top 100 companies were highly profitable. Among individual firms, the top-ranked Acer had higher annual sales than any other firm but had a moderate 12.6 percent of ROA relative to its sales. Other large-scale manufacturers such as Acer Peripherals, GVC, and First International also had relatively moderate performances in terms of ROA and total sales in 1995. In contrast, Asustek Computer Inc. and Kuo Feng Corp. both had very impressive ROAs as well as low debt/asset ratios.

Figure 8.2 illustrates the profitability of eighty-two of Taiwan's top one hundred electronic parts and components manufacturers. The top-ranked (Taiwan Philips Electronics Industries) and the third-ranked (Philips Electronic Building Elements Industries Taiwan Ltd.) manufacturers in this category were not included in this analysis because of data unavailability. The three top-five electronic components manufacturers (Tatung, TI, and Matsushita) did not perform very well financially; all of them had a below-industry-average ROA. In contrast, many lower-ranked firms, such as ISSI (ranked 59), Sun Plus (96), and Etron (98), performed remarkably well. In addition, Etron and Sun Plus both have a very low debt/asset ratio. Other well-known semiconductor manufacturers, such as TI-Acer, UMC, TSMC, and Winbond, had healthy performances during 1995.

In general, Taiwanese electronics firms' profitability, measured by corporate accounting rate of return, did not increase with market share, at least according to the 1995 data. This implies that the achievement of the minimum efficient scale of production for most manufacturers does not require a very large market share to gain economies of scale. Thus, a firm's profitability depends more upon industry characteristics and corporate strategies than upon the leader's share of industry sales.

Overall, the top hundred electronic parts and components manufacturers were much more profitable (with an average ROA of about 10 percent) than the downstream computer and communication equipment makers in 1994 and 1995. This phenomenon reflects the lower entry barrier and price competition of the computer and communications segments of the industry, and also reflects the general industrial trend of shifting the higher value-added activities toward the upstream parts and components business in the industry.

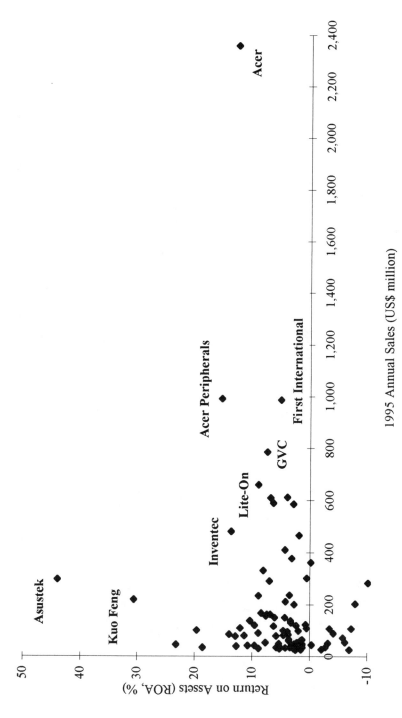

Figure 8.1 Profitability of 89 of Taiwan's Top 100 Information and Communication Firms

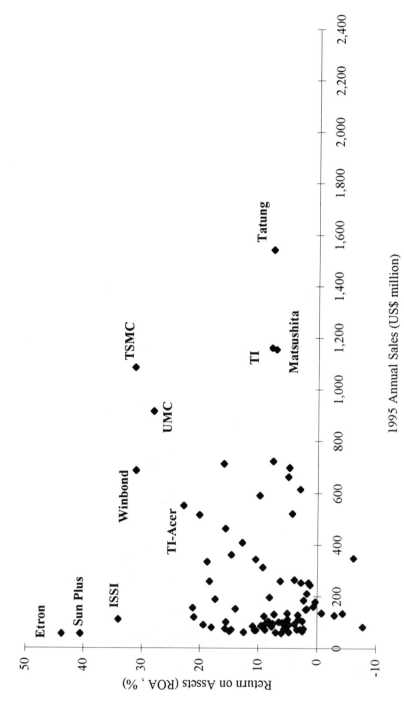

Figure 8.2 Profitability of 82 of Taiwan's Top 100 Electronic Parts & Components Firms

8.3. Industry Classification and Competitive Strategies

Except for the beverage industry, the Taiwanese electronic components industry as a whole had the highest rate of profitability in 1995, while the information products and communications segments had the fastest rate of growth in sales. Table 8.1 illustrates the inter-industry performance of Taiwan's top 1000 manufacturing firms.

Table 8.1 Performances of Taiwan's Top 1000 Manufacturers in 1995 by Industry Categories (Units: US$ million)

Industry	No. of firms	Total sales $	Avg. profit $	Avg. sales $	Profitability* (%)	Sales growth (%)
Electronics (parts & components)	182	31,133	26.04	170.94	15.4	29.4
Information/ communications	102	19,031	8.68	186.42	4.7	43.1
Apparel	15	769	0.75	51.32	2.1	5.8
Automobile/ parts	39	9,553	1.51	244.91	0.7	8.0
Beverage	9	4,584	355.85	509.43	69.9	-2.1
Chemical engineering products	31	1,645	2.26	53.21	4.5	12.1
Chemical materials	50	10,328	20.75	206.42	10.2	26.7
Detergents/ cosmetics	12	1,340	4.53	111.70	4.1	18.9
Education/ entertainment	2	144	-16.43	72.08	-23.0	42.8
Food/ animal feed	77	9,445	8.30	122.64	7.1	12.0
Forest/ranch	2	144	-16.23	72.08	-23.0	24.8
Machine tools/ equipment	53	3,564	4.91	67.17	7.7	9.5
Metal products	51	3,126	3.77	61.13	6.3	13.9

Table 8.1 (Cont.)

Non-metal mineral products	36	4,471	16.60	124.15	13.7	0.2
Other transportation equipments	23	3,343	8.30	145.28	5.9	4.1
Paper/printing	23	3.247	7.17	141.13	5.2	24.1
Petroleum/ coal products	2	11,752	272.45	5,876.23	4.6	11.9
Pharmaceutical industry	16	1,033	3.02	64.53	5.2	12.9
Plastics	27	5,326	12.45	197.36	6.3	16.6
Precision instruments	7	563	1.89	80.38	2.7	17.4
Rubber	11	1,108	4.53	100.75	4.9	8.6
Shoes/ leather	18	1,322	3.02	73.58	4.3	6.4
Steel/ basic metal	100	14,078	6.79	140.75	4.9	22.3
Textile	94	12,607	7.92	133.96	6.1	18.5
Wood/ non-metal products	13	685	3.02	52.83	5.9	-6.4
Others	5	214	3.77	43.02	8.8	8.7

* Profit after taxes/sales
Source: Taiwan Common Wealth Magazine, June 5, 1996. (US$1 = NT$26.5)

As Table 8.1 and Figures 8.1 and 8.2 indicate, there is as much difference in profitability within the industry as among various industries. Profit differences within industries are important, because they reflect the fact that competitors pursue different strategies within the same industry, with differing levels of efficiency. Profit differences within industries also depend on industry-level attributes, such as the threat of new entrants, bargaining power of buyers and suppliers, the threat of substitutes, and the intensity of rivalry within the industry [Porter, 1980]. Different industries afford competitors different opportunities to make commitments of long-lived resources. Such commitments play a critical role in sustaining profit differentials among industry incumbents and between incumbents and potential entrants [Ghmawat, 1991].

One way to understand the size and sources of such differences is to classify industries in terms of the resources (e.g., capacity, customer base, or knowledge) that dominate competition [Collis and Ghemawat, 1994]. Profit differences within knowledge-driven industries usually reflect differences in competitors' abilities to develop significant innovations. Effective manufacturing strategies in the dynamic electronics industry depends on intensive R&D efforts to continuously develop new product and process technologies. However, large capital requirements in the electronics industry, especially the semiconductor fabrication facilities (fabs) which cost over US$ 50 million per fab, also serve as a major barrier to competition. In the face of growing global competition, more capital-intensive strategies are now being pursued by both private firms (e.g., Formosa Plastics and its subsidiaries) and the government in Taiwan.

In general, industrial strategies in newly industrialized countries (NICs) electronics industries are characterized by the transformation of firms from OEMs (original equipment manufacturers) to ODMs (original design manufacturers), and ultimately to establish their own product brand names in order to become OBMs (original brand manufacturers) such as firms that fully integrate the functions of design, manufacturing, and sales of their own brands under registered trademarks.

OEMs produce proprietary products to their OBM customers' design specifications. By adding assembly and test capabilities, many OEM firms that began as parts and components vendors in Taiwan now produce products for their OBM customers. This is necessary because their main customers may find a more attractive alternative source, especially given the rising labor cost and because the profit margins on subcontracting are so low. Growing competition from low-cost competitors is requiring Taiwanese, as well as other NIC firms to increase their involvement in the product design process. Innovative design capability provides the basis for introducing higher value-added products. Over time, as OEM firms acquire and increase their design capabilities, they gradually become ODM firms. Future ODM capability is dependent upon the availability of highly trained and experienced design engineers. The base of engineering graduates in Taiwan is the largest among the Asian NICs. This abundance of strategic human resources provides Taiwan a distinct competitive advantage over its global competitors.

The two largest notebook computer makers, Inventec and GVC, have become 100 percent ODM producers selling products under their customers' brand names. For example, Inventec produces Apple's Newton PDA and provides Compaq with its high-end LTE 5000 series notebook computer. GVC is the original design manufacturer for Packard Bell. The availability of "ready to go" products gives the ODM competitor a definite advantage over the traditional OEM (e.g., Tatung and Nan-Ya) producer in building close relationships with global customers.

The majority of OEM and ODM suppliers do not have recognizable brand names or direct access to markets or end users. The lack of recognized brands makes new entry into established markets difficult. Therefore, it is important to develop globally recognized names as well as market and related distribution channels. In the near future, as design capabilities and marketing skills become mature, more and more Taiwanese electronics manufacturers will gradually move from being ODM makers to OBM producers. Taiwan's largest desktop and notebook computer maker, Acer, has already become one of the well-known OBM producers in the global market. Inventec, also is applying its technology towards becoming the largest OBM producer of hand-held electronic dictionaries under its own "Besta" brand name in the Chinese market.

Customer service is also one of the strategic requirements for global success. Most Taiwanese OEM or ODM manufacturers provide warranties through the national service centers of the brand holder. The capability of rapid response to customer problems and the ability to identify potential problems before the final products are shipped become important competitive factors. Effective communication to correct the source of failures and the development of quality assurance programs are central to the industries' long-term success.

Another growing industrial trend in the 1990s is the large amount of overseas investment by Taiwanese electronics firms, especially among the information electronics manufacturers. Low-cost manufacturing, local market knowledge, closeness to the market, as well as a strong visible presence are regarded as a source of global competitive advantage. The decision to establish manufacturing plants overseas can be explained by "pull" and/or "push" factors [Taggart and McDermott, 1993]. Taiwanese electronics manufacturers have been "pulled" toward Europe and North America by the size of the market and protectionism, but "pushed" by high wage increases to locate labor-intensive production initially in Malaysia and Thailand, and recently in Chinese mainland and the Philippines to take advantage of low-cost labor.

The establishment of many overseas manufacturing and assembling facilities has enabled many Taiwanese firms to explore new business opportunities, such as the implementation of global logistics systems (GLS). GLS is the hottest subject in 1996 among Taiwanese information electronics firms. This is because the global computer giant IBM decided to transfer part of its worldwide business of PC manufacturing, delivering, as well as after-sales services to Taiwanese firms.

GLS is different from the traditional OEM or ODM. When implementing a GLS with global computer giants, Taiwanese manufacturers not only are responsible for the designing and manufacturing of components, but also are responsible for assembling and delivering the final products to the prime contractors' customers, as well as performing the after-sales services. The rewards for the implementation of GLS are certainly higher than the traditional OEM or ODM services. However, the management of global material flows and inventory, as well as the allocation of financial resources will add additional cost burdens to the subconstracting manufacturers. Nevertheless, many Taiwanese information electronics firms, such as First International, GVC, and Elitegroup, have aggressively explored such new business opportunities. Acer already decided to implement its own GLS in the early 1997 through its 37 assembling locations throughout the world to assemble, delivery, and service its brand name products.

Chapter 9

CURRENT STATUS AND FUTURE DIRECTIONS

Overall, the current state and future outlook of Taiwan's electronics industry, which includes computers, communications, consumer electronics, and semiconductors, is strong. This chapter provides an assessment of the current status and future growth of Taiwan's electronics industry and discusses several key factors necessary for success.

9.1 Taiwan's Information Electronics Industry

Taiwan's information product and technology industry has the following key characteristics:

- Taiwan's information industry, especially PCs, will continue to grow substantially due to the global economic recovery. Table 9.1 shows the double-digit growth rates of Taiwan's information industry between 1994 and 1995. The trend is projected to continue over the next several years.

- Mergers, acquisitions, and alliances among domestic manufacturers and product market diversification is growing. Larger corporations with diversified product and market strategies will dominate the smaller ownership/partnership single-product companies.

- Because production costs are higher in Taiwan than in the Philippines and China, the production of low-end products will be gradually transferred to those countries. Close-to-market manufacturing is another trend among Taiwanese information electronics producers. Mexico and Great Britain usually are the places considered to set up manufacturing facilities due to their proximity to growing markets. Taiwan's information industry will be emphasizing the manufacturing of high-value-added products, and will expand its research and development centers in those areas.

113

Table 9.1 Taiwan Information Industry Growth Outlook (Unit: US$ million)

	1994	1995	Growth rate(%)
Hardware product value	11,579	13,139	13.5
Information services	1,431	1,693	21.0
Overseas product value	3,003	4,279	42.4

Exchange rate: 26.42(1994) and 26.50 (1995).
Source: Taiwan Institute for Information Industry, 1995

- In addition to Acer Corp., several other Taiwanese information electronics manufacturers have started to establish their own images in the global marketplace. The trend toward converting Taiwan's information product manufacturers from OEMs to ODMs and name-brand producers and OBMs has already begun.

- The personal computer has changed from a simple personal productivity improvement tool to a ubiquitous, versatile, multimedia product that can be used for a variety of purposes throughout industry and commerce. In 1996 and beyond, the top selling products will still be multimedia (e.g., Pentium and PCI-bus personal computers, 17-inch monitors, CD-ROM drives, MPEG, and related multimedia chipsets), and network-related products, such as high-speed communications products.

In the past, the capability of releasing new products quickly, pricing products competitively, and turning out superior products has made Taiwan's information electronics industry more competitive than its counterparts in South Korea and other countries. However, Taiwanese firms must undergo several fundamental changes in the fast changing global competitive environment in order to maintain and strengthen their global competitiveness in the information electronics market.

Several global-oriented strategies, some already implemented by many Taiwanese information electronics firms, will assist Taiwanese firms to maintain high rates of growth well into the next century. These strategies include:

- Forming strategic alliances with global electronics firms as part of their efforts at reconfiguration and coordination of global value chain activities. Participating in global alliances will help Taiwanese firms to open new market opportunities, reduce risk and direct competition, and combine competencies to enhance competitive advantages.

- Strengthening industry global competitive capability by establishing a complete global system of division of labor, and sufficiently taking advantage of different areas of competitive advantage to increase industry global competitiveness. For example, First International's US$ 100 million project to establish global material flows and logistics systems to provide its OEM constractors instant supports and services. In addition to the eight existing manufacturing and service centers located in the United States, mainland China, Germany, and Czech Republic, First International plans to establish another twenty or more PC assembling plants and service centers in Brazil, Mexico, and Europe by the year of 2000.

- Enhancing global intelligence systems to closely monitor worldwide product market trends and technology developments.

- Reducing the reliance on foreign competitors for the supply of key materials, components, and manufacturing equipment by building vertical integration capabilities and/or working closely with the upstream electronics industry. Other efforts, such as acquiring key product and process technologies through international joint ventures or technology exchange activities, should also be intensified.

- Enhancing software and services as a key to the information electronics industry in order to increase global competitiveness.

9.2. Prospects for the Taiwan's Consumer Electronics Industry

Since the 1970s, East Asia has become the global center for the production of consumer electronics. In the 1980s, the production of color TVS, car radios and audio stereos, and videotape-recorders increased rapidly in Taiwan and South Korea. However, for the past decade, due to the strong Taiwan dollar, higher labor costs, limited industrial land, and competition from other NICs and China, the total values of production and export have drastically decreased.

The decline of the traditional consumer electronics industry indicates that this product market has already entered a new transition period. Taiwan's consumer electronics industry will be undertaking fundamental changes in order to survive the next round of global competition. Strategies for future development of Taiwan's consumer electronics industry will include

- product digitalization;
- development of key components of electronics products;
- improvement of conventional electronics manufacturing technology;
- product quality improvement, marketing and image promotion;
- human resource development to improve consumer electronics design;
- promotion of a heathy competitive environment; and
- strengthening the structure of global division of labor.

9.3. Taiwan's Electronics Components Industry

Since the early 1990s, several major changes have occurred in Taiwan's electronics components industry. First, firms are increasing investment and expanding facilities optimistically and aggressively. Second, several major electronics firms (e.g., UMC and TSMC) are beginning to participate in the high-value-added market for strategic IC components, such as CPU and DRAM, which they had previously shied away from due to inadequate capabilities. Finally, non-volatile memory, a specially designed technology, has increasingly become a hotly pursued product in Taiwan's semiconductor market.

Overall, the potential for future development of Taiwan's electronics components industry remains excellent. Looking back to past developments and examining the current status of Taiwan's semiconductor industry reveals several advantages.

- Strong capability to learn, develop, and diffuse new technology: Since ERSO first transferred MOS manufacturing process technology (7μ) from RCA in 1976, Taiwan developed a complete technological infrastructure for the electronics industry. Much of the IC manufacturing and design technology and core technical team were transferred from ERSO. UMC was the first IC manufacturer spin-off from ERSO in 1980. This was followed by TSMC (1987), Winbond Electronics (1987), Hualon Microelectronics (1987), and IC design company spin-offs, including Syntek (1982), Princeton (1986), SIS (1987), and Proton Electronic Industrial Co., Ltd. (1985). The strong ability to learn and develop new technologies reflects the excellent quality of technical personnel in Taiwan. Highly educated personnel continually enter the emerging high-technology industries. The strong ability to diffuse technology with high payoffs has promoted capital investments.

- Taiwan's electronics industry incorporates many visionary entrepreneurs and excellent business management personnel. Many Taiwanese entrepreneurs were able to identify and take advantage of unique opportunities to invest in the electronics industry. For example, UMC' and TSMC's success depended on the vision and management skills of their founders.

- A strong downstream personal computer industry supports the upstream research and technology development. Taiwan's companies are perhaps currently less capable in either design or process technology than American and Japanese firms. However, Taiwanese firms do possess both countries' strengths in many product areas. In the memory IC area, such as SRAM, Taiwan has UMC and Winbond; in the chipset category, Taiwanese manufacturers such as SIS, ALI, UMC, and VIA all have balanced technology and product developments. This unique characteristic has made it very difficult for South Korea -- Taiwan's chief competitor - to compete in the global electronics or semiconductor markets. In addition, Taiwan has a strong downstream PC industry to support the further development of the semiconductor industry; this is a strength that many of Taiwan's competitors do not possess.

• The industry has easy access to a low-cost capital supply from Taiwan's financial market. Taiwan's semiconductor manufacturers have been riding a wave of economic expansion since 1990; consequently, most of them have accumulated a large amount of capital over the years, and are ready for reinvestment. The diversity of Taiwan's capital market also contributed to the development of the electronics industry. Publicly traded electronics companies can easily obtain cheap capital for investment, reducing the cost of operation. Moreover, firms are able to control their own investment opportunities.

• High-quality human resources are sufficient to develop high-technology industries. Over the years, Taiwan has sent thousands of students to study abroad, especially in the United States. After they became scientists or engineers, many returned to Taiwan and contributed to the high-technology development of Taiwan's electronics industry. Recently, after the establishment of HSIP, more overseas well-educated Chinese have returned to Taiwan (most in the field of electrical engineering and computer science). This "reverse brain drain" is expected to continue, and further strengthen the technological and industrial capabilities of Taiwan. In addition to the return of overseas Chinese scientists and engineers, a large number of science and engineering graduates are coming out of the local universities every year. This large pool of high quality human resources provide a solid foundation for future development of high-tech industries in Taiwan.

9.4. Training and Education

Since the 1970s the main thrust of Taiwan's educational policy has been directed to the training of electrical engineers for the domestic information and semiconductor industry. Since the shortage of engineers and technically skilled labor had grown into a major bottleneck for these industries, special training centers were created outside of the university system. The "Computer Communicating Laboratory" (CCL), under ITRI's purview, offers hardware programs and III software programs. The Office for Industrial Development, on its own initiative, provides basic and pre-university level curricula for the information industry.

Continuing education in the private sector is very poor. As a result, the III established two training centers, one each for the commercial labor force and for the technical management force. Special continuing education courses are offered for the semiconductor skilled workforce in Hsinchu science park. In fiscal year 1992-93, those programs enrolled 1,620, down from 2,061 in 1991-92.

9.5 R&D Support

In terms of electronics technology acquisition, transfer, and dissemination, the government's ERSO has played the lead role in Taiwan. Most companies having industrial and technological potential are located in the Hsinchu science park, 80 km south of Taipei. Without its steady support, Taiwan's semiconductor industry would probably not have been able to develop to its present effectiveness. In four domestic research projects, technological competence was gradually developed in all phases of the IC production process and passed on to private industry via spin-offs or through other channels (see Figure 3.2). The heavy support for design process is remarkable. It was the DRAM shortage in the late eighties that first brought about a greater concentration on the capability for integration and kicked off the current "sub-micron" project [Liu, 1993]. ERSO's continued commitment to the 64 MB DRAM/16 MB SRAM technology is currently not planned; efforts are to be autonomously continued by the industry which has grown to industrial maturity (ERSO shouldered the total costs of the first three projects). In the "sub-micron" project, industry, represented by UMC and TSMC (plus a symbolic contribution from Etron, Holtek, Macronix, Mosel-Vitelic, and Winbond), is sharing for the first time half of the current costs; these companies have received low-interest loans for this purpose. ERSO is making its laboratory and a hundred development scientists available, and is financing the acquisition of equipment and devices.

At present, alongside the "sub-micron" project are a number of other domestic research projects targeting industrial development of semiconductor technology or applied technologies, under the overall leadership of ERSO or other institutions.

9.6. Hsinchu Science-based Industrial Park

As discussed earlier, the Hsinchu science-based industrial park (HSIP) was established in 1980 to encourage the development of technology and the creation of high-technology companies by Taiwan. While there are some bio-technology firms within the park, the primary focus of the HSIP is on the electronics industry. Today, all of Taiwan's semiconductor industries are located there; the semiconductor industry comprises 24 percent of the park's companies and 37 percent of the park's turnover. Other sectoral foci are the EDV industry, telecommunications, and optoelectronics. Only companies that meet specific criteria of technological intensity and sales profitability are allowed to set up and stay there.

The park, which is administered by a division of the NSC, the Science-based Industrial Park Administration, has put Taiwan on the world map of IC manufacturing and key information technology industry components. IC manufacturing has been aided by a full range of support industries that handle materials, design, testing, and packaging. Profit margins averaging nearly 25 percent in recent years have allowed companies located in the park to spend between 5 and 6 percent of revenues on research and development.

Among the listed companies based in the science park are TSMC and UMC, Taiwan's two largest semiconductor chip makers, as well as Mosel-Vitelic. The Hsinchu science-based industrial park, set up nearly fifteen years ago as the incubator for high-tech, has been one of Taiwan's greatest industrial-policy success stories. More than 180 companies located there are likely to have combined sales of about US$ 10 billion in 1995, about 50 percent more than last year, according to Director-General Steve Hsieh [Bloomberg, 1995].

9.7. Reversed Brain Drain

In the early years of economic development, while the government was actively engaged in promoting the quality of its human resources, Taiwan was encountering a "brain-drain" problem; that is, a number of those who received higher education have gone abroad for advanced studies and never returned. This is particularly true for students in the fields related to science and industrial technology. In response to the brain-drain problem, the government in Taiwan has adopted various policies to attract those who study abroad to return home to work and live in Taiwan. However, the preferential treatment to attract overseas Chinese scholars has not proved to be very effective.

As Taiwan rapidly transformed herself into a newly industrialized economy, the brain-drain problem seems to have improved significantly. The most fundamental factor behind the rapid improvement was not government policy, but the strong demand for high-level manpower in the private sector. A kind of reverse brain drain has occurred in Taiwan area in recent years. More students are staying at home for higher education and more are returning from overseas to seek work in Taiwan because of limited job opportunities in the West. The government encourages students educated overseas to return by offering them a financial incentive from the National Youth Commission under the Executive Yuan upon their return.

Before 1980, less than one thousand overseas graduates came home to seek employment each year. The annual figure jumped to 5,000 in 1992 and is growing, Meanwhile, the number of students receiving diplomas from Taiwan's own colleges and universities is growing by about 33,000 per year. In 1992, there were 30,000 students in postgraduate schools around the island, double the 1987 total. The result is a growing pool of highly educated persons competing for top positions.

9.8. Final Assessment

The development of Taiwan's industrial capabilities has been and is being supported in Taiwan by comprehensively designed endeavors on the R&D, infrastructure development, and investment levels. It may be conjectured that, among the East Asian countries, the relative scope and intensity of Taiwan's industry support are the highest and its relationship with China may become the strongest benefit to both countries.

The industrial policy support of the domestic electronics industry was designed for the long term from the outset and effectively and successfully shaped industrial development. The success of Taiwan's electronics industry is based on three factors: first, the generally exceptional conditions for industrial production; second, the steady "reverse brain drain" from Silicon Valley; and, third, the demand pull of Taiwan's electronic data processing (EDP) industry. Industrial policy has developed under these conditions and helped a vibrant, typically mid-sized industrial policy activities, the national research projects for development of a domestic technological competency and liberal support for investment and production are likely to have been the most significant.

Whereas in Japan industrial policy is pursued through rather close cooperation between the private sector and government and in Korea the economic policy climate presides with the chaebols, in Taiwan overall industrial policy management clearly resides with governmental authorities. The initiative for originating and developing a domestic electronics industry in Taiwan emerged from government agencies, especially from the MOEA, which embarked upon development of ERSO in the early seventies under the leadership of Minister Y.S. Sun. The decision-making and opinion-building process was and is done openly and transparently, and includes foreign experts.

Unlike Korea and Japan, Taiwan is able to implement industrial policy free from foreign trade constraints and pressures. There are a number of reasons for that. Access to the market in Taiwan is unhindered; foreign companies may participate in the growing domestic market by means of imports, direct investments, and technology transfer. Another advantage is that the strategy for catching up industrially has not concentrated on single products that subsequently clog the global market. Instead, the industrial policy is designed for broad technological dissemination, global integration, and import substitution. This results in very few points of friction with overseas firms.

The future prospects for Taiwan's electronics industry are excellent. The indicated strengths (cost and sizing advantages, international networking, growing domestic market, flexibility and dynamism of the mid-size sector) should prove sustainable over the medium to long haul. Taiwan's companies will continue their successful strategy; that is, on the one hand, nipping at the heels of pioneering technological developments with a justified R&D outlay and, on the other hand, focusing on the production of smaller batch sizes. In this way it is likely to continue to be present in the global market in a timely fashion with technologically competitive products.

LIST OF ACRONYMS

AMD	Advanced Micro Devices, Inc.
AMLCD	active matrix liquid crystal display
APSI	Assistance Program for Strategic Industries
ASIC	application-specific integrated circuit
BGA	ball grid array
CCL	Computer Communicating Laboratory
CD-ROM	compact disc read only memory
CEPD	Council for Economic Planning and Development
CETRA	China External Trade Development Council
COG	chip-on-glass
CPE	customer premises equipment
CRT	cathode ray tube
DIP	dual inline package
DRAM	dynamic random access memory
EDP	electronic data processing
EPROM	erasable programmable read only memory
ERSO	Electronics Research and Service Organization
FDI	foreign direct investment
FPD	flat panel display
GI	General Instrument
GLS	global logistics system
HDTV	high definition television
HSIP	Hsinchu Science-based Industrial Park
IC	integrated circuit
III	Institute for the Information Industry
IMI	International Micro Industries
ISSI	Integrated Silicon Solution (Taiwan) Inc.
ITIS	Industrial Technology Information Services
ITRI	Industrial Technology Research Institute
IUS	Industrial Upgrading Statute
LCD	liquid crystal display
LED	light emitting diode
MCM	multi-chip modules
MOEA	Ministry of Economic Affairs

MOS	metal oxide semiconductor
MPEG	motion picture expert group
NIC	newly industrialized country
NII	National Information Infrastructure
NIST	National Institute of Standard and Technology (U.S. Department of Commerce)
NSC	National Science Council
OEM	original equipment manufacturer
ODM	original design manufacturer
OBM	original brand manufacturer
PDA	personal digital assistant
PGA	pin grid array
PLCC	plastic leaded chip carrier
PMLCD	passive matrix liquid crystal display
PQFP	plastic quad flat pack
PTH	plated through hole
R&D	research and development
RISC	reduced instruction set computer
ROA	return on assets
S&T	science and technology
SCR	silicon-control rectifier
SEI	Statute for Encouragement of Investment
SIS	Silicon Integrated System Corp.
SME	small and medium enterprise
SMT	service mount technology
SOP	small outline package
SRAM	static random access memory
STN	super twisted nematic
TFT	thin film transistor
TI	Texas Instruments
TN	twisted nematic
TSMC	Taiwan Semiconductor Manufacturing Corp.
UEC	Union Electronics Corp.
UMC	United Microelectronics Corp.
VLSI	very large scale integration
WTV	widescreen television

REFERENCES

Balassa, B. (1988). The Lessons of East Asian Development: An Overview, *Economic Development and Culture Change*, 36(1), pp. 149-167.

Bloomberg, (1995). Science Park-based Companies Facing Loss of Tax Breaks, *China News*, December 16, p.5.

Chang, P., Shin, C. and Hsu, C. (1993). Linking Technology Development to Commercial Application, *International Journal of Technology Management*, 8(6/7/8), pp. 697-712.

Chang, P., Shih, C. and Hsu, C. (1994). The Formation Process of Taiwan's IC Industry - Method of Technology Transfer, *Technovation*, 14(3), pp. 161-171.

Chen, S. (1995). Memory IC Products Case Study, Taipei, Taiwan: Industrial Technology Information Services, (in Chinese).

Chen, H. (1994). IC Market Application Analysis, Taipei, Taiwan: Industrial Technology Information Services, (in Chinese).

CNA (1995). Lee Salutes 15 Years of Science Park, *China News*, December 16, p.3.

Collis, D. and Ghemawat, B. (1994). Industry Analysis: Understanding Industry Structure and Dynamics, in Fahey, L. and Robert, M.R. (eds.) *The Portable MBA in Strategy*, New York, NY: John Wiley & Sons, Inc., pp. 171-194.

Costlow, T. and Patterson, A. (1993). Taiwan, U.S. To Build Infrastructure; Tech Exchange Bolsters MCMs; MCM infrastructure Building In U.S., Taiwan, *EE Times,* November 29, p.1 and 86.

Culpan, R. (1993). Multinational Competition and Cooperation: Theory and Practice, in Culpan, R. (ed.) *Multinational Strategic Alliances,* Binghamton, NY: International Business Press, pp. 13-32.

Ghemawat, P. (1991). *Commitment: The Dynamic of Strategy*, New York, NY: The Free Press.

Golden, W.T. (ed.) (1991). *Worldwide Science and Technology Advice - to the Highest Levels of Governments*, New York: Pergamon Press.

Hagge, J.K. and Wager, R.J. (1995). Known Good Die Workshop in 44th and 45 ECTC, May 21.

Hobday, M. (1995). Innovation in East Asia: Diversity and Development, *Technovation*, **15**(2), pp. 55-63.

Hou, C. and Chang, C. (1981). Education and Economic Growth in Taiwan: The Mechanism of Adjustment." in *Conference on Experiences and Lessons of Economic Development in Taiwan*, 492. Taipei: The Institute of Economics, Academia Sinica.

Hou, C. and San, G. (1993). National Systems Supporting Technical Advance in Industry: The case of Taiwan," in Nelson, R. R. (ed.) *National Innovation Systems -- A Comparative Analysis*, New York, NY: Oxford University Press, pp. 384-413.

Hou, C. and Wu, H. (1983). Wage and Labor Productivity in the ROC." in *Raising Productivity: Experience of the Republic of China*, Tokyo: Asian Productivity Organization.

Hsiao, F. (1995). *Taiwan's Industrial Development Strategy toward the 21st Century*, Taipei, Taiwan: Council for Economic Planning & Development, Executive Yuan, (in Chinese).

Hwang, C. (1995). *Taiwan - the Republic of Computers*, Taipei, Taiwan: Commonwealth Publishing Co., Ltd., (in Chinese).

IEEE Special Report (1991). "Asiapower," *IEEE Spectrum*, June, pp 24-67.

Li, K.T. (1991). The Role of Foreign Science Advisers in the Republic of China (Taiwan), in Golden, W.T. (ed.) *Worldwide Science and Technology Advice*, New York, NY: Pergamon Press, pp. 133-143.

Liu, C. (1993). Government's Role in Developing a High-Tech Industry: The Case of Taiwan's Semiconductor Industry, *Technovation*, **13**(5), pp. 299-309.

Nelson, R.R. (ed.) (1993). *National Innovation Systems - A Comparative Analysis*, New York, NY: Oxford University Press.

Patterson, A., (1994). Partners with Taiwan Group In SMART DIE Effort; Intel Forges MCM Links, *EE Times*, July 11, p. 18.

Porter, M.E. (1980). *Competitive Strategy*, New York, NY: The Free Press.

Chiang, M. (1994). *IC Packaging Products Case Study*, Taipei, Taiwan: Industrial Technology Information Services, (in Chinese).

Science Park Administration, (1995). *Science-Based Industrial Park*, Hsinchu, Taiwan, March, (in Chinese).

Soesastro, H. and Pangestu, M. (eds.) (1990). *Technological Challenge in the Asia-Pacific Economy*, Winchester, MA: Unwin Hyman Inc.

Solid State Technology, (1996). World News, April, p.36.

Taggart, J.H. and McDermott, M.C. (1993). *The Essence of International Business*, New York: NY: Prentice-Hall.

Taiwan Industrial Technology Research Institute, (1995). *1995 Taiwan Semiconductor Yearbook*, Hsinchu, Taiwan, (in Chinese).

Taiwan Industrial Technology Research Institute, (1996). *1996 Taiwan Semiconductor Yearbook*, Hsinchu, Taiwan, (in Chinese).

The Economist, (1992). A Survey of Taiwan - A Change of Face, October 10.

The Republic of China Yearbook 1996, (1996). Taipei, Taiwan: Government Information Office.

The World Bank, (1993). *The East Asian Miracle - Economic Growth and Public Policy*, New York, NY: Oxford University Press.

U.S. Congress, Office of Technology Assessment, (1995). *Flat Panel Displays in Perspective*, OTA-ITC-631, Washington, D.C.: U.S. Government Printing Office, September.

World Journal, (1996a). High-Speed Digital Communication Operations will be Opened to Private Enterprises, July 31, (in Chinese).

World Journal, (1996b). Changing Strategies for Overseas Manufacturing Bases of Taiwan's Information Electronics Industry, July 31, (in Chinese).

Wu, S. (1992). The Dynamic Cooperation Between Government and Enterprise: The Development of Taiwan's Integrated Circuit Industry, in Wang, N.T. (ed.)

126

Taiwan's Enterprises in Global Perspective, Armonk, NY: M.E. Sharpe, Inc., pp. 171-192.

Yoshino, M.Y. and Rangan, U.S. (1995). *Strategic Alliances - An Entrepreneurial Approach to Globalization*, Boston, MA: Harvard Business School Press.

INDEX

Acer, 25, 33, 36, 41, 53, 54, 60, 69, 71, 77, 84, 87, 89, 106, 107, 112, 114
AMD, 41, 88
APSI, 18
ASIC, 41, 73-74, 77, 93
AST, 33

BGA, 81, 97, 98

CD-ROM, 43, 51-52, 59, 114
China, 1, 3, 5, 7, 11, 15, 42, 50, 55, 58, 64, 67, 70, 83, 95-96, 105, 113, 115, 119
China External Trade Development Council (CETRA), 3
Chunghwa Picture Tubes, 59, 100, 102
COG, 100
Compaq, 33, 50, 53, 83, 87, 111
Council for Economic Planning and Development (CEPD), 12, 28
CPE, 55-56
CRT, 59, 99-102
Cyrix, 41

Dell, 33
DIP, 81-82

electronic data processing (EDP), 42, 120
Electronics Research and Service Organization (ERSO), 20-21, 35-36, 38-42, 44-45, 97-98, 110, 116, 118, 119, 120
Elitegroup Computer Systems, 43, 35, 112
Etron, 37, 69, 71, 90, 106, 108, 118
EPROM, 88, 90-91

First International, 46, 60, 106, 112, 115
foreign direct investment (FDI), 18, 26
Formosa Plastics Corp., 46,111
foundry, 68-69, 73-75, 77-78, 83, 88

General Instruments (GI), 38, 94-95
global logistics system (GLS), 112
GVC, 53, 54, 106-107 111-112

HDTV, 59-60
Hitachi, 60, 86, 93-94, 100
Hong Kong, 1, 4, 9-10, 64, 66, 81, 83
Hsinchu Science-based Industrial Park (HSIP), 27-28, 87, 103, 117-118
Hualon, 38, 40-41, 73, 116

IBM, 28, 33, 53, 84, 87, 105, 112
Industrial Technology Research Institute (ITRI), 4, 13, 19-20, 22-23, 25, 27, 31, 35, 38-40, 43-45, 54, 81, 100, 102, 117
Industrial Upgrading Statute (IUS), 19
Institute for the Information Industry

Appendix A: 1995 Top 150 Electronic Parts & Components Companies in Taiwan (US$ Million)

Rank	COMPANY (ELECTRONIC PARTS & COMPONENTS)	1995 Sales	Sales growth (%)	Assets	Profit after taxes	Net worth	Capital	Debt/asset %	Employee Number
1	PHILIPS ELECTRONIC BUILDING ELEMENTS INDUSTRIES (TAIWAN) LTD.	1,691.92	64.94	N/A	N/A	N/A	163.17	-	7,993
2	TATUNG CO.	1,542.23	12.98	2,929.25	222.15	1,408.04	N/A	51.93	11,105
3	PHILIPS ELECTRONIC BUILDING ELEMENTS INDUSTRIES (TAIWAN) LTD.	1,433.85	35.86	N/A	N/A	N/A	46.04	-	3,828
4	TEXAS INSTRUMENTS TAIWAN LTD.	1,161.06	47.93	449.25	35.36	205.47	28.23	54.26	2,625
5	MATSUSHITA ELECTRIC (TAIWAN) CO., LTD.	1,156.08	11.90	733.21	52.49	333.66	101.17	54.49	5,481
6	TAIWAN SEMICONDUCTOR MFG. CO., LTD. (TSMC)	1,085.51	48.76	1,822.75	569.09	1,267.13	543.02	30.48	3,412
7	UNITED MICROELECTRONICS CORP. (UMC)	914.98	59.07	1,806.64	507.21	1,202.49	507.09	33.43	2,982
8	CHUNGHWA PICTURE TUBE, LTD.	874.72	32.92	1,360.53	N/A	N/A	347.13	-	5,543
9	TECO ELECTRIC & MACHINERY CO., LTD.	723.28	12.33	923.32	70.00	504.30	242.49	45.38	3,593
10	MOTOROLA ELECTRONICS TAIWAN LTD.	713.32	26.78	296.68	47.32	240.87	38.83	18.81	2,800
11	GENERAL INSTRUMENT OF TAIWAN LTD.	699.02	13.88	266.38	12.91	94.00	46.64	64.71	4,500
12	WINBOND ELECTRONIC CORP.	687.55	113.97	1,118.45	347.66	661.81	204.72	40.82	2,352
13	SAMPO CORP.	664.60	10.44	575.55	28.87	324.34	N/A	43.64	3,645
14	PACIFIC ELECTRIC WIRE & CABLE CO., LTD.	616.30	13.14	1,627.77	48.26	653.55	481.92	59.85	1,874
15	WALSIN INC.	591.02	22.75	1,539.09	151.43	1,020.83	544.53	33.67	1,555
16	TEXAS INSTRUMENTS - ACER INC.	551.47	73.50	1,075.02	245.40	526.11	261.51	51.06	1,405
17	KINPO ELECTRONICS INC.	520.64	11.93	307.70	13.32	156.11	90.68	49.26	1,650
18	MOSEL VITELIC INC.	515.28	103.80	1,412.45	283.62	948.49	283.02	32.84	1,757

Rank	COMPANY (ELECTRONIC PARTS & COMPONENTS)	1995 Sales	Sales growth (%)	Assets	Profit after taxes	Net worth	Capital	Debt/asset %	Employee Number
19	DELTA ELECTRONICS INC.	462.34	35.26	312.04	48.75	212.49	112.38	31.90	3,040
20	HON HAI PRECISION INDUSTRY CO., LTD.	407.77	63.45	357.77	45.66	205.77	86.00	42.48	1,025
21	TAIWAN HITACHI CO., LTD.	389.17	12.56	274.34	N/A	N/A	49.32	-	1,260
22	ADVANCE SEMICONDUCTOR ENGINEERING INC.	361.81	94.64	600.75	88.04	411.36	150.87	31.52	3,387
23	CHUNG-HSIN ELECTRIC & MACHINERY MFG. CORP.	349.17	-9.61	608.34	-37.36	202.42	154.72	66.72	1,793
24	TAIWAN LITON ELECTRONIC CO., LTD.	344.38	42.43	229.58	24.04	157.13	91.21	31.55	1,122
25	MACRONIX INTERNATIONAL CO., LTD.	334.57	61.87	620.26	116.49	420.57	188.68	32.19	1,754
26	SHIHLIN ELECTRIC & ENGINEERING CORP.	313.43	7.87	357.89	33.21	199.85	106.60	44.15	2,180
27	TAIWAN FUTABA ELECTRONICS CORP.	295.62	13.12	153.09	N/A	N/A	55.66	-	2,000
28	HUA ENG WIRE & CABLE CO., LTD.	264.98	21.80	359.55	13.89	268.19	175.96	25.40	604
29	SANYO ELECTRIC (TAIWAN) CO., LTD.	259.81	12.79	317.28	19.96	173.77	102.38	45.23	2,214
30	MICROCHIP TECHNOLOGY TAIWAN	258.57	65.74	142.23	26.11	78.30	13.96	44.94	446
31	TAI-I ELECTRIC WIRE & CABLE CO., LTD.	254.19	19.85	309.70	8.60	114.91	41.51	62.89	500
32	SHINLEE CORPORATION	254.08	0.62	257.25	3.85	90.60	55.36	64.77	1,072
33	KAOHSIUNG HITACHI ELECTRONICS CO., LTD.	244.79	26.40	98.57	1.21	23.92	12.68	75.72	1,938
34	TDK ELECTRONIC TAIWAN CORP.	242.75	7.43	159.70	N/A	69.74	12.30	56.33	1,884
35	TAIWAN KOLIN CO., LTD.	211.89	-9.85	350.79	6.19	169.17	127.81	51.77	1,496
36	UNIVERSAL SCIENTIFIC INDUSTRIAL CO., LTD.	196.53	36.37	148.34	12.00	61.77	42.30	58.35	1,234
37	COMPEQ MANUFACTURING CO., LTD.	189.58	38.55	257.58	44.72	156.64	63.77	39.18	1,676
38	TA YA ELECTRIC WIRE & CABLE CO., LTD.	185.40	24.00	314.04	7.21	178.11	128.19	43.28	717
39	ACER GLOBAL INC.	179.55	58500.00	56.23	0.19	3.81	3.77	93.22	157

Rank	COMPANY (ELECTRONIC PARTS & COMPONENTS)	1995 Sales	Sales growth (%)	Assets	Profit after taxes	Net worth	Capital	Debt/asset %	Employee Number
40	HITACHI TELEVISION (TAIWAN) LTD.	166.42	39.46	48.49	N/A	17.85	11.32	63.19	670
41	PACIFIC GLASS CORP.	161.81	15.08	143.70	N/A	N/A	48.08	-	499
42	TAIWAN FLUORESCENT LAMP CO., LTD.	160.75	3.07	274.00	1.74	150.94	81.51	44.91	1,481
43	SILICON INTEGRATED SYSTEMS CORP.	155.58	58.09	143.66	30.45	112.38	N/A	21.77	275
44	SILICON WARE PRECISION INDUSTRIES CO., LTD.	152.38	59.98	337.70	46.91	289.25	105.92	14.33	1,720
45	SINOCA ENTERPRISES CO., LTD.	151.89	108.87	142.72	2.49	51.43	37.74	63.96	1,000
46	SONY VIDEO TAIWAN CO., LTD.	149.09	14.09	34.87	0.68	15.85	8.75	54.54	499
47	PHILIPS TAIWAN LTD.	144.45	9.65	N/A	N/A	N/A	7.55	-	528
48	SANYO ELECTRONIC (TAICHUNG) CO., LTD.	141.40	20.13	54.64	N/A	N/A	10.11	-	825
49	TAIWAN TOYO RADIO CO., LTD.	135.74	-0.19	24.68	-0.19	12.79	7.55	48.16	1,024
50	TAIAN ELECTRIC CO., LTD.	135.06	19.26	129.66	6.57	62.00	32.19	52.15	759
51	GREAT CHINA ELECTRIC WIRE & CABLE CO., LTD.	134.72	25.83	58.72	-2.53	14.38	17.32	75.51	80
52	CHINA ELECTRIC MFG. CORP.	131.40	13.08	169.43	12.38	122.60	79.85	27.63	1,035
53	TSANG KUEN ENTERPRISE CO., LTD.	129.09	-4.73	143.36	4.64	31.13	13.13	78.25	280
54	EFA CORP.	127.74	13.82	78.00	-2.26	28.49	22.83	63.47	664
55	SOLOMON TECHNOLOGY CORP.	123.17	43.22	92.64	8.34	40.94	27.58	55.80	474
56	WUS PRINTED CIRCUIT CO., LTD.	120.11	56.56	104.26	21.85	76.49	46.68	26.63	1,308
57	GLORIA ELECTRONICS CO., LTD.	117.06	16.52	98.68	N/A	37.43	26.42	62.10	525
58	PICVUE ELECTRONICS LTD.	111.70	17.69	248.98	12.42	126.98	97.47	49.00	1,556
59	INTEGRATED SILICON SOLUTION (TAIWAN) INC.	111.21	108.71	87.77	29.85	52.34	18.79	40.37	170
60	ACTION ELECTRONICS CO., LTD.	110.75	16.05	74.91	2.34	29.70	22.60	60.30	402

Rank	COMPANY (ELECTRONIC PARTS & COMPONENTS)	1995 Sales	Sales growth (%)	Assets	Profit after taxes	Net worth	Capital	Debt/asset %	Employee Number
61	PROTON ELECTRONIC INDUSTRIAL CO., LTD.	105.02	36.22	101.51	2.53	41.02	N/A	59.55	792
62	CHING KUAG CHEMICAL CO., LTD.	103.58	19.97	115.62	9.51	55.21	44.42	52.25	400
63	SILITEK CORP.	101.47	46.69	149.85	9.74	97.51	58.87	34.92	548
64	PHOENIXTEC POWER CO., LTD.	101.47	35.12	101.32	15.70	81.13	45.28	19.92	468
65	TAI HONG CIRCUIT INDUSTRIAL CO., LTD.	99.51	22.36	75.62	4.45	35.55	13.21	52.99	600
66	MABUCHI MOTOR CO. LTD.	97.09	18.46	69.36	6.00	52.60	18.53	24.15	1,835
67	FORWARD ELECTRONICS CO., LTD.	96.91	13.83	57.85	N/A	33.28	15.09	42.40	1,141
68	DIALER & BUSINESS ELECTRONICS CO., LTD.	95.62	35.94	147.89	11.09	102.45	52.38	30.72	425
69	TAIWAN SAKURA CORP.	93.77	18.55	154.49	5.96	95.89	61.89	37.93	977
70	HUALON MICROELECTRONICS CORP.	91.02	1.51	284.23	13.66	204.75	N/A	27.94	1,039
71	HOLTEK MICROELECTRONICS INC.	90.45	31.12	126.72	24.57	82.04	50.75	35.25	990
72	JUNG SHING WIRE CO., LTD.	87.89	29.60	63.28	6.00	37.40	20.68	40.90	466
73	DEE VAN ENTERPRISE CO., LTD.	87.02	2.90	56.00	5.32	26.30	15.09	53.03	356
74	YAGEO CORP.	84.79	39.65	365.25	39.77	310.26	101.89	15.05	912
75	ZTONG YEE INDUSTRIAL CO., LTD.	83.36	8.97	106.38	-8.26	15.17	22.72	85.74	1,100
76	HONG TAI ELECTRIC INDUSTRIAL CO., LTD.	82.87	3.92	100.91	7.77	81.02	54.34	19.70	320
77	EPSON INDUSTRIAL (TAIWAN) CORP.	82.30	11.90	45.92	N/A	19.85	11.81	56.77	880
78	TAIWAN YAZAKI CORP.	80.94	-21.88	N/A	N/A	N/A	N/A	-	1,316
79	GOLD CIRCUIT ELECTRONICS LTD.	80.38	102.66	67.62	12.15	28.49	11.32	57.86	534
80	OPTO TECH CORP.	75.55	15.32	94.53	2.15	46.57	29.43	50.73	511
81	MOLEX TAIWAN LTD.	75.51	44.89	58.45	9.13	43.25	19.55	26.01	450

Rank	COMPANY (ELECTRONIC PARTS & COMPONENTS)	1995 Sales	Sales growth (%)	Assets	Profit after taxes	Net worth	Capital	Debt/asset %	Employee Number
82	DAH SAN ELECTRIC WIRE & CABLE CORP.	75.25	28.64	53.25	2.98	23.13	17.02	56.55	327
83	KAIMEI ELECTRONIC CORP.	73.51	13.25	57.58	1.66	19.92	15.85	65.40	610
84	TAIWAN TENLON CO., LTD.	72.49	-6.10	35.47	N/A	N/A	N/A	-	1,200
85	ALLIS ELECTRIC CO., LTD.	72.23	23.00	86.64	2.94	34.04	25.92	60.71	486
86	UNICAP ELECTRONICS INDUSTRIAL CORP.	71.62	44.99	73.21	10.68	42.42	28.64	42.06	792
87	WORLD WISER ELECTRONICS INC.	70.34	134.76	88.57	7.81	53.40	29.92	39.71	869
88	CHIN-POON INDUSTRIAL CO., LTD.	69.74	24.19	58.23	6.15	33.77	20.94	41.99	649
89	ULTIMA ELECTRONIC CORP.	69.55	23.36	53.43	5.55	24.11	16.98	54.87	300
90	PIONEER ELECTRONIC (TAIWAN) CORP.	65.89	24.00	30.34	0.75	15.09	5.40	50.24	560
91	LIEN CHANG ELECTRONIC ENTERPRISE CO., LTD.	65.43	31.06	40.26	6.00	26.04	11.40	35.33	290
92	AMBIT MICROSYSTEMS CORP.	62.42	279.35	35.17	4.38	14.08	13.43	59.97	314
93	TEAPO ELECTRONIC CORP.	61.40	3.17	59.51	2.87	28.34	21.47	52.37	691
94	FOSTER ELECTRIC CO.,(TAIWAN) LTD.	60.42	11.72	19.92	1.40	6.53	3.58	67.23	419
95	FORTUNE ELECTRIC CO., LTD.	57.62	13.27	67.17	4.08	33.92	25.06	49.49	364
96	SUN PLUS	55.74	32.34	45.36	18.38	37.28	14.94	17.80	103
97	CENTRAL INDUSTRY CO., LTD.	55.55	78.20	51.92	N/A	1.81	0.94	96.51	228
98	ETRON TECHNOLOGY INC.	55.13	107.23	45.09	19.74	33.92	11.51	24.68	81
99	SINO-AMERICAN ELECTRONIC CO., LTD.	55.06	8.79	29.02	N/A	13.74	6.98	52.66	399
100	COROWAN TAIWAN CORP.	55.06	19.68	20.49	1.21	8.45	3.77	58.74	280
101	PRINTED WIRE CORP.	54.23	36.85	61.92	6.75	33.89	22.23	45.27	520
102	(WEI-SUN)	54.00	-	33.85	3.51	18.83	N/A	44.37	130

Rank	COMPANY (ELECTRONIC PARTS & COMPONENTS)	1995 Sales	Sales growth (%)	Assets	Profit after taxes	Net worth	Capital	Debt/asset %	Employee Number
103	HUA JUNG COMPONENTS CO., LTD	53.62	30.72	57.25	3.66	26.72	15.09	53.32	735
104	ESSEX MONITOR CO., LTD.	53.51	-16.34	N/A	N/A	N/A	N/A	-	200
105	CHANT WORLD INTERNATIONAL CO., LTD.	53.02	-1.12	82.26	3.85	63.55	28.94	22.75	702
106	TAIWAN MURATA ELECTRONICS CO., LTD.	52.83	8.78	N/A	N/A	N/A	10.19	-	300
107	PHILIPS LIGHTNY TAIWAN LTD.	51.58	22.49	N/A	N/A	N/A	7.55	-	135
108	YAHSIN INDUSTRIAL CO., LTD.	50.83	39.58	55.77	N/A	14.38	11.51	74.22	400
109	SKYNET ELECTRONIC CO., LTD.	48.64	16.75	25.96	N/A	10.23	7.51	60.61	415
110	HUNTER FAN TAIWAN CO.	47.13	29.16	5.89	0.11	0.79	0.19	86.53	124
111	TAIWAN THICK-FILM IND CORP	47.06	36.43	37.81	2.72	12.08	5.21	68.06	312
112	CHINA SEMICONDUCTOR CORP.	46.87	47.15	52.79	-1.89	20.34	25.77	61.47	434
113	TA HSING ELECTRIC WIRE AND CABLE CO., LTD.	46.75	9.84	82.15	0.34	31.09	14.45	62.15	335
114	TERA ELECTRONIC CO., LTD.	46.75	59.66	131.43	12.11	121.21	N/A	7.78	88
115	LONGWELL CO.	45.89	11.97	40.38	1.17	23.96	16.60	40.65	390
116	YUNG FU ELECTRICAL APPLIANCES CORP., LTD.	45.58	2.54	85.92	0.53	11.51	5.66	86.60	295
117	DYNA IMAGE CORP.	45.32	39.16	33.51	5.28	19.28	13.13	42.45	427
118	ELITE MATERIAL CO., LTD.	44.72	65.96	59.47	6.23	28.15	24.15	52.66	181
119	EVERTOP WIRE CABLE CO., LTD.	44.68	56.82	40.68	4.00	18.60	11.36	54.26	168
120	PAN INTERNATIONAL INDUSTRIAL CORP.	43.51	32.37	74.49	2.26	46.30	28.98	37.84	750
121	ELAN MICRO-ELECTRONICS CORPORATIONS	43.43	496.37	63.36	4.26	41.32	37.74	34.78	129
122	SIRTEC INTERNATIONAL CO., LTD.	43.36	16.17	46.79	0.75	14.49	12.23	69.03	260
123	REALTEK SEMI-CONDUCTOR CO., LTD.	43.13	30.77	33.70	3.74	21.51	15.81	36.17	217

Rank	COMPANY (ELECTRONIC PARTS & COMPONENTS)	1995 Sales	Sales growth (%)	Assets	Profit after taxes	Net worth	Capital	Debt/asset %	Employee Number
124	LINGSEN PRECISION INDUSTRIES, LTD.	42.75	43.59	56.00	3.28	37.09	19.40	33.76	739
125	HANPIN ELECTRON CO., LTD.	42.34	1.63	70.53	0.91	26.45	15.13	62.49	310
126	KAOLUNG LUNG TAMURA ELECTRONICS CO., LTD.	40.45	2.68	18.19	-1.21	6.34	7.96	65.14	287
127	WONDERFUL WIRE CABLE CO., LTD.	40.11	11.77	49.06	1.70	23.62	16.98	51.84	276
128	EVERLIGHT ELECTRONICS CO., LTD.	39.81	43.73	41.51	4.04	24.94	13.21	39.90	469
129	RECTRON LTD.	39.74	55.31	53.36	-9.43	29.13	35.92	45.40	300
130	CAESAR	39.51	21.18	85.17	1.77	41.66	31.13	51.08	600
131	TAIWAN CHEMI-CON CORPORATION	38.98	13.26	24.94	N/A	19.28	15.09	22.69	249
132	HI-SINCERITY MICROELECTRONICS CORP.	38.83	55.67	49.47	3.70	20.83	14.94	57.89	592
133	KUANG YUAN CO., LTD.	37.47	24.43	18.94	1.77	8.87	7.17	53.18	308
134	ASE TEST, INC.	36.83	94.81	75.85	15.70	46.94	23.55	38.10	611
135	YETI ELECTRONICS CO., LTD.	36.26	17.48	29.74	4.83	14.98	8.57	49.61	237
136	MERRY ELECTRONICS CO., LTD.	35.47	0.10	33.81	3.09	18.57	12.72	45.08	N.A.
137	SANKO ELECTRIC CO., LTD.	35.21	12.13	13.74	N/A	6.00	2.26	56.31	330
138	HUA CHENG ELECTRIC CO., LTD.	35.02	-2.62	17.17	N/A	14.42	5.74	15.82	324
139	UNI-CIRCUIT CO., LTD.	34.38	61.81	27.70	5.25	10.42	4.79	62.39	299
140	MMC ELECTRONICS TAIWAN CO., LTD.	34.26	7.20	N/A	N/A	N/A	6.34	-	410
141	TRANDON ELECTRONIC INDUSTRIAL CO., LTD.	34.23	-4.52	31.70	0.75	17.66	N/A	44.16	274
142	VERTEX PRECISION ELECTRONICS, INC.	34.11	31.39	31.47	4.30	16.49	11.32	47.60	263
143	CHANTEK ELECTRONIC CO., LTD.	34.08	45.17	81.17	5.81	44.15	28.68	45.60	553
144	SUPER ELECTRONICS CO., LTD.	34.00	2.03	36.00	N/A	29.58	7.40	17.82	380

Rank	COMPANY (ELECTRONIC PARTS & COMPONENTS)	1995 Sales	Sales growth (%)	Assets	Profit after taxes	Net worth	Capital	Debt/asset %	Employee Number
145	POTRANS ELECTRICAL CORP., LTD.	34.00	34.88	24.94	3.09	14.68	11.51	41.15	309
146	BOARDTEK ELECTRONICS CORP.	33.81	3.94	29.62	5.32	15.58	11.32	47.38	287
147	QUALITEK ELECTRONICS CO., LTD.	33.55	16.51	N/A	N/A	N/A	6.79	-	310
148	LOGITECH INC.	33.32	108.25	13.47	-0.87	8.26	N/A	38.65	587
149	LU CHIANG ELECTRIC WIRE & CABLE INC.	33.13	27.98	20.04	0.38	9.70	N/A	51.60	78
150	EPISIL TECHNOLOGIES INC.	32.91	65.77	33.13	3.89	19.25	15.09	41.91	302

Appendix B: Taiwan's Major Electronic Parts & Components Product Manufacturers

Company	Founded	Location	President	Contact	Phone	Fax	1	2	3	4	5	6	7	8	9
Acer	1987	Taipei	C.J. Wu	K.L. Pan	02-762-8800	02-762-6138	x								
ADT	1981	HSIP	W.K. Wu	W.K. Wu	03-577-3120	03-577-3123			x						
Advanced Test	N/A	HSIP	C. Tsai	S. Guo	03-578-8780	03-579-9560							x		
AMPI	1987	HSIP	H.Y. Cheng	H.Y. Cheng	03-577-0030	03-577-8211				x					
Analog Interactions	1992	HSIP	M.J. Lee	S.F. Lai	03-577-2500	03-577-2510	x								
Aplus	1992	Taipei	S.T. Kuo	S.Y. Su	02-781-8277	02-781-5779	x								
ASE	1983	Kaohsiung	C.H. Chang	C.H. Chang	07-361-7131	07-361-3094					x				
ASE Test Inc.	1988	Kaohsiung	S.H. Pan	S.H. Pan	07-363-6641	07-361-5446						x			
Aslic Micro Electronics Corp.	1987	Taipei	C.S. Chiu	Ramon Tsai	02-501-4996	02-505-6421	x								
AT&T Taiwan Telecommunication	N/A	Taipei	N/A	W.F. Cheng	02-547-3131	02-547-3131	x								
ATC	1983	HSIP	S.Y. Wu	Y.M. Yen	03-577-7300	03-577-6464									x
AVID	1995	HSIP	G. Chen	C.M.Lou	03-579-5222	03-578-7789	x								
Cadence	1985	HSIP	C.K. Xu	H.C. Chiao	03-577-8951	03-577-6257					x				
Caesar Technology Inc.	1991	Hsinchu	Charles Liu	C.M. Lin	035-824500	035-824847				x					
Campass	N/A	Taipei	C.L. Peng	C.C. Chang	02-345-5170	02-757-6013					x				
Chantek Electronics Corpl	1989	Hsinchu	Y.H. Tsien	S.S. Cheng	03-551-9181	03-551-9088				x					
Chesen	1984	Taipei	S.F. Chang	S.F. Chang	02-916-1299	02-910-4935					x				
CHIM	1995	Taipei	Y.M. Chang	Y.J. Lai	02-346-0539	02-346-0850	x								
China Seminconductor Corp. (CSC)	N/A	Taipei	Johnson Chen	Ben Yang	02-788-1818	02-788-3332							x		
Chino-Excel Technology Corp.	N/A	Taipei	Y.H. Lin	Sunny Fu	02-537-6305	02-537-6305				x					

*Product Category

Company	Founded	Location	President	Contact	Phone	Fax	Product Category*									
							1	2	3	4	5	6	7	8	9
Chip Design	1985	Taipei	S.T. Lin	P.F. Guo	02-299-4908	02-299-0133	x								
CSC	1973	Taipei	W.C. Tsai	B.Y. Yang	02-223-9696	02-223-9377								x	x
E-Cmos	1988	HSIP	C.C. Hu	C.W. Chen	03-578-3622	03-578-3630		x							
ELAN	1994	HSIP	Y.H. Yeh	C.S. Wong	03-578-7505	03-577-9095		x							
E-Team	1992	Hsinchu	M.T. Lu	C.C. Teng	03-572-8535	03-572-8526						x			
Episil Technologies Inc.	1985	HSIP	M.C. Huang	L.W. Yin	03-5779246	035-776289	x			x					
EPLUS	N/A	Taipei	C.C. Wang	Y.T. Chen	02-754-8038	02-706-3617		x							
Etron Technology Inc.	1991	HSIP	C.C. Lu	G.W. Cheng	03-578-2345	03-579-9001		x							
Everlight Electronics Co., Ltd.	N/A	Taipei	Robert Yeh	Kent Cheng	02-260-2000	02-260-6189									x
First International Computer Inc.	1993	HSIP	Ming Chien	W.K. Yang	03-578-1314	03-577-5314					x		x		
Frontier Electronics Co., Ltd.	1984	Taipei	Daniel Huang	P.H. Kao	02-914-7685	02-911-9143								x	
General Instrument of Taiwan Ltd.	N/A	Taipei	Ronald Ostertah	R. Ostertah	02-911-3861	02-917-5991								x	
Ginjet	1989	Taipei	M.D. Lin	M.D. Lin	02-506-3439	02-507-2064		x							
Hi-Sincerity Microelectronics Corp.	1984	Taipei	C.L. Xu	Andy Wu	02-521-2056	02-563-2712								x	
Hitachi	1987	Taipei	S.H. Chen	C.D. Chou	02-718-3666	02-718-8180		x		x					
Holtek Microelectronics Inc.	1988	HSIP	C.Y. Wu	Jonny Wong	03-578-4888	03-577-0879		x							
Holylite Microelectronics	1992	Hsinchu	S.D. Lin	S.D. Lin	03-542-0523	03-542-1634		x							
Hualon Microelectronics Corp.	1987	HSIP	D.M. Wong	T.Y. Tieng	03-577-4945	03-577-4305				x					
Innova	1989	HSIP	T.C. Peng	Y.H. Huang	03-578-2366	03-578-2019			x						
Integrated Silicon Solution, Inc. (ISSI)	1990	HSIP	K.Y. Han	Y.C. Chen	03-578-0333	03-578-3000		x							
Keystone	N/A	Taipei	W.L. Huang	H.M. Yio	02-392-2124	02-321-1272		x							

Company	Founded	Location	President	Contact	Phone	Fax	1	2	3	4	5	6	7	8	9
Kingbright Electronic Co., Ltd	1980	Taipei	W.B. Soong	N/A	02-249-9224	02-240-3981									x
Ledtech Electronics Corp.	1977	Taipei	Frank Liu	Frank Liu	02-218-6891	02-218-6222									x
Lingsen Precision Industries Ltd.	1973	Taichung	K.C. Yeh	K.C. Yeh	04-533-5120	04-532-7904					x				x
LITEON	1975	Taipei	Raymond Soong	N/A	02-222-6181	02-221-0660									x
LSI	1989	Taipei	C.L. Bai	M.C. Liu	02-718-7828	02-718-8869	x								
Macronix International Co., Ltd.	1989	HSIP	M.C. Wu	W.L. Chien	03-578-8888	03-578-8887			x						
MEMC	1994	HSIP	N/A	K.L. Tsai	03-578-3131	03-578-7287	x								
Mentor	N/A	Taipei	P.K. Dai	N/A	02-757-6050	02-757-6027						x			
Micor Advance	1992	Taipei	Y.W. Mao	S.Y. Chang	02-760-8850	02-762-6099	x								
Micro Electronics	1991	Taipei	W.B. Lin	W.B. Lin	02-999-1822	02-999-4991	x								
Micro Silicon	1989	Hsinchu	P.T. Chang	N/A	035-965770	035-953278					x				
Microchip Technology Taiwan	1996	Kaohsiung	M. Hao	S.H. Chang	07-821-2171	07-831-1593					x				
More Power	1989	Hsinchu	P.T. Cheng	C.H. Peng	03-595-9213	03-595-5611					x				
Mos Design Semiconductor Corp.	1988	Taipei	W.H. Chu	D.C. Chen	02-918-2105	02-913-9312	x								
Mosart	1993	Taipei	C.S. Jong	C.S. Jong	02-959-9180	02-959-9323	x								
Mosel-Vitelic Inc.	1991	HSIP	H.C. Hu	Y.P. Cheng	03-577-0055	03-577-2595			x						
Mospec Semiconductor Corp.	1987	Tainan	C.Y. Kao	James Wu	06-599-1621	06-599-1626							x		
Motorola Electronics Taiwan Ltd.	N/A	Taipei	N/A	N/A	02-717-7089	02-717-5047	x				x				
Myson	1991	HSIP	Y.F. Tang	K.L. Su	03-578-4866	03-578-5002	x								
Nan-Ya	1995	Taipei	Y.S. Chuang	S.H. Wang	02-712-2211	02-719-7413			x						
National Semiconductor	1979	Taipei	T.S. Cheng	S.C. Cheng	02-521-3288	02-561-3054	x								

Company	Founded	Location	President	Contact	Phone	Fax	Product Category* 1	2	3	4	5	6	7	8	9
Opto Tech Corporation	1983	HSIP	M.D. Lin	S. Shyu	03-577-7481	03-577-9576									x
Orient	N/A	Kaohsiung	G.Y. Tu	W.F. Fu	07-361-3131	07-363-2319					x				
Philips Elec. Building Elem. Industries	N/A	Kaohsiung	K.C. Huang	K.C. Huang	07-361-2511	07-361-2164					x				
Philips Electronics Ind. (Taiwan)	N/A	Taipei	Y.C. Lo	J.S. Lin	02-382-4763	02-382-4777	x								
Photron Semiconductor Corp.	N/A	Hsinchu	Allen Tan	N/A	03-598-5889	03-598-2700								x	
Powerchip	1994	HSIP	T.J. Huang	K.C. Tsai	03-578-7899	03-578-8565				x					
Princeton Technology Corp.	1986	Taipei	Richard Chiang	H.C. Pan	02-917-5004	02-917-4598	x								
Progate	1991	Taipei	C.S. Lai	C.S.Lai	02-759-0680	02-759-0408	x								
Prosperity Electronics Co., Ltd.	N/A	Taoyuan	W.C. Huang	W.C. Huang	03-388-5216	03-388-5219								x	
Realtek	1987	HSIP	J.G. Huang	R. Lin	03-578-0211	03-577-6047	x								
Rectron Ltd.	1976	Taipei	Y.T. Wang	K.Y. Sun	02-758-9155	02-268-7815								x	
Roco	N/A	Taipei	C.B. Lin	W.B. Tsai	02-766-0156	02-761-5691	x								
Samsung	N/A	Taipei	N/A	C. Chen	02-757-7292	02-757-7311	x								
Sanyo Electric (Taichung) Co., Ltd.	1976	Taichung	N/A	N/A	04-532-3141	04-533-0041					x				
SARC	N/A	Taipei	T.N. Huang	W.C. Shyu	02-726-6460	02-759-3174	x								
Silicon Integrated Systems Corp. (SIS)	1987	HSIP	C.Y. Duh	Ken Huang	03-577-4922	03-577-8774	x								
Silicon Ware	1993	HSIP	W.P. Lin	Y.H. Huang	03-534-1525	03-533-0639					x		x		
Siliconix	1974	Kaohsiung	Charles Ku	F.T. Shien	07-361-5101	07-361-3484					x			x	
Siliconware Precision Industries Co.	1984	Taichung	C.L. Lin	C.L. Lin	04-534-1525	04-533-0639					x				
Silitek Corporation	N/A	Taipei	Raymond Soong	George Cheng	02-432-4123	02-432-4924								x	
Sino-American Silicon Products, Inc.	1981	HSIP	Davie Yen	Y. Lou	03-577-2233	03-578-1706	x								

Company	Founded	Location	President	Contact	Phone	Fax	1	2	3	4	5	6	7	8	9
Sinonarch	1988	HSIP	C.S. Lee	T.K. Lee	03-578-3366	03-578-1812									x
Sun Plus	1990	Hsinchu	J.C. Huang	I.H. Chen	03-578-6005	03-5786006	x								
Syntek Design Technology Co., Ltd.	1983	Taipei	David Wang	C.M. Lee	02-505-6383	02-506-4323	x								
Taicera	1979	Taichung	W.P. Lin	W.P. Lin	04-532-2111	04-533-7146					x				
Taiwan Mask Corp.	1988	Hsinchu	B.W. Cheng	K.N. Shien	03-578-1370	03-578-0752			x						
Taiwan Memory	1993	Hsinchu	H. Chiao	S. C. Lu	03-578-7720	03-578-7719	x								
Taiwan Semiconductor Co. Ltd.	1979	Taipei	Arthur Wang	Arthur Wang	02-917-4145	02-912-2499								x	
Tamarack	1987	Taipei	P.S. Huang	W.S. O-Young	02-773-3086	02-776-0545	x								
Tatung Co.	N/A	Taipei	T.S. Lin	A.C. Wang	02-592-5252	02-592-5185	x								
Texas Instruments Taiwan Ltd.	N/A	Taipei	H.P. Lo	C.L. Kuo	02-378-6800	02-377-2718	x								
Texas Instruments Taiwan Ltd.	N/A	Taipei	J. Ker	C.L. Guo	02-943-5141	02-314-0994					x				
TI-Acer Inc.	1990	HSIP	S.C. Chen	S.C. Chen	03-578-5112	03-578-2038				x					
Tontek Design Technology Co., Ltd.	1986	Taipei	Sam Hsiau	S.C. Huang	02-222-4475	02-222-4764	x								
TSMC	1987	HSIP	Brooks	K.C. Lin	03-578-0221	03-578-1546				x					
Tyntek	1988	HSIP	M.C. Shyu	M.C. Shyu	03-578-1616	03-578-0545									x
UMC	1980	HSIP	M.C. Shuang	T.J. Wang	03-577-3131	03-577-0584				x					
USC	1995	HSIP	Y.D. Liu	N/A	03-579-5158	03-579-5166				x					
Utron	1993	HSIP	C.C. Kuo	C.C. Kuo	03-577-7882	03-577-7919	x								
V-Tac	1992	Taipei	C.F. Chiu	W.H. Lin	02-796-2880	02-781-5690	x		x						
Vanguard	1994	HSIP	F.C. Tseng	C.J. Chien	03-577-0355	03-578-1920			x						
VATE	1988	HSIP	F.C. Shih	F.C. Shih	03-577-0345	03-577-0668							x		

Company	Founded	Location	President	Contact	Phone	Fax	Product Category*								
							1	2	3	4	5	6	7	8	9
VIA	1987	Taipei	W.C. Cheng	F.L. Chu	02-218-5452	02-218-5453	x								
VLSI	1990	Taipei	H.L. Lo	L.H.Chang	02-719-5466	02-718-3204	x								
Walsin Inc.	N/A	Taipei	Y.H. Chiao	S.Y. Liu	02-712-7734	02-715-3383					x			x	x
Weltrend Semiconductor, Inc.	1989	HSIP	S.M. Lin	T.F. Tsai	03-578-0241	03-577-0419	x								
Winbond Electronics Corp.	1987	HSIP	T.Y. Yang	M.H. Liao	03-577-0066	03-579-2668			x						
Yuban	N/A	Taipei	C.L. Chao	C.H.Tsai	02-773-0022	02-731-2698	x								

*Product Category: **1.** Wafer Material, **2.** IC Design, **3.** Mask, **4.** IC Manufacture, **5.** Packaging, **6.** Design Tools, **7.** Test, **8.** Discrete Components, **9.** Optoelectronics.

Appendix C: 1995 Top 100 Information Electronics and Communications Products Companies in Taiwan (US$ million)

Rank	Company	1995 Sales	Sales growth (%)	Assets	Profit after taxes	Net worth	Capital	Debt/Assets %	Employee Number
1	ACER INC.	2,359.13	89.05	1,657.47	208.91	1,004.60	356.64	39.38	3,788
2	ACER PERIPHERALS INC.	992.87	67.83	338.91	51.89	164.94	71.70	51.33	1,478
3	FIRST INTERNATIONAL COMPUTER INC.	987.28	57.39	692.26	35.13	305.55	150.04	55.86	2,600
4	GVC CORP.	785.36	50.99	433.66	32.34	276.72	98.57	36.18	1,870
5	LITE-ON TECHNOLOGY CORP.	660.34	97.57	265.66	23.89	82.45	46.49	68.96	1,011
6	DIGITAL EQUIPMENT INTERNATIONAL LTD. (TAIWAN)	612.08	13.82	163.77	6.45	85.92	18.23	47.53	1,610
7	MAG TECHNOLOGY CO., LTD.	609.06	43.56	328.53	22.68	110.26	67.28	66.43	1,357
8	ADVANCED DATUM INFORMATION CORP.	589.96	40.40	487.09	31.06	243.96	N/A	49.91	1,418
9	CHUNTEX ELECTRONIC CO., LTD.	586.72	57.48	345.25	9.96	148.34	96.98	57.03	1,504
10	INVENTEC CORP.	481.47	81.95	327.21	44.79	141.36	65.74	56.79	1,945
11	MITAC INTERNATIONAL CORP.	466.49	63.62	304.19	5.81	158.98	107.66	47.73	1,135
12	COMPAL ELECTRONICS INC.	411.89	12.61	248.15	10.83	179.77	97.47	27.54	1,449
13	TWINHEAD INTERNATIONAL CORP.	378.23	108.50	145.74	4.57	41.06	37.32	71.82	1,096
14	ELITEGROUP COMPUTER SYSTEMS CO., LTD.	361.77	58.75	242.34	-0.45	106.87	55.02	55.88	1,200
15	QUANTA COMPUTER INC.	330.72	9.08	120.64	9.77	57.81	N/A	52.08	900
16	JEAN CO., LTD.	301.32	140.22	130.91	0.72	25.25	25.02	80.71	2,462
17	ASUSTEK COMPUTER INC.	297.06	134.14	167.17	73.58	121.62	22.64	27.24	570
18	CLEVO CO.	290.04	28.67	100.49	7.09	43.36	30.91	56.85	826
19	CHICONY ELECTRONICS CO., LTD.	283.13	50.24	129.77	-13.13	46.64	56.60	64.05	680
20	TAIWAN INTERNATIONAL STANDARD ELECTRONICS LTD.	268.26	-4.29	220.26	N/A	137.89	16.60	37.39	1,228

Rank	Company	1995 Sales	Sales growth (%)	Assets	Profit after taxes	Net worth	Capital	Debt/Assets %	Employee Number
21	CAPETRONIC (KAOHSIUNG) CORP.	243.17	27.95	60.79	N/A	N/A	20.45	-	1,104
22	VIDAR-SMS CO., LTD.	236.75	34.34	292.91	10.30	117.89	103.77	59.75	1,133
23	ORIENT SEMICONDUCTOR ELECTRONICS LTD.	234.42	91.49	206.87	18.53	115.21	62.15	44.30	2,265
24	KUO FENG CORP.	218.87	8.37	250.42	76.53	164.68	40.94	34.23	700
25	SAMPO TECHNOLOGY CORP.	212.23	8.00	98.30	4.15	41.43	31.70	57.85	808
26	TECO INFORMATION SYSTEM CO., LTD.	202.72	5.33	121.92	-9.66	29.92	45.28	75.45	890
27	ROYAL INFORMATION ELECTRONICS CO., LTD.	201.25	8.10	154.68	4.23	46.53	33.96	69.91	514
28	PRIMAX ELECTRONICS LTD.	167.06	19.71	105.36	8.91	67.40	37.32	36.03	753
29	AT&T TAIWAN TELECOMMUNICATIONS CO., LTD.	162.98	-9.28	178.72	12.34	79.62	58.79	55.44	872
30	MICROTEK INTERNATIONAL INC.	162.75	44.00	191.58	14.34	108.30	78.75	43.47	711
31	SHAMROCK TECHNOLOGY CO., LTD.	156.30	62.81	66.23	5.06	23.89	16.45	63.87	620
32	ASE TECHNOLOGIES INC.	149.89	32.40	97.66	4.26	22.23	22.64	77.24	715
33	BEHAVIOR TECH COMPUTER CORP.	149.62	4.15	155.62	9.55	63.81	29.02	58.99	879
34	UMAX DATA SYSTEMS INC.	137.70	70.83	119.28	12.38	55.92	33.96	53.11	643
35	WYSE TECHNOLOGY TAIWAN LTD.	137.32	22.48	197.40	6.53	167.02	99.77	15.38	855
36	CIS TECHNOLOGY INC.	130.04	136.02	170.57	5.58	65.96	39.62	61.32	990
37	CHAPLET SYSTEMS INC.	123.06	7.80	105.09	0.75	13.36	11.32	87.25	400
38	D-LINK CORP.	120.19	30.90	122.15	11.70	82.68	40.00	32.31	835
39	SIEMENS TELECOMMUNICATION SYSTEMS LTD.	119.13	-37.34	187.17	4.42	129.17	37.74	30.98	666
40	ASKEY	116.98	83.75	69.06	4.34	23.70	13.25	65.68	450
41	GIGA-BYTE TECHNOLOGY CO., LTD.	110.00	107.17	43.66	5.28	10.26	3.62	76.49	190

Rank	Company	1995 Sales	Sales growth (%)	Assets	Profit after taxes	Net worth	Capital	Debt/Assets %	Employee Number
42	DATAEXPERT CORP.	108.26	8.10	64.08	0.34	18.87	15.09	70.61	334
43	AQUARIUS SYSTEM INC.	107.25	-1.14	65.85	-4.79	21.51	N/A	67.39	400
44	AOC INTERNATIONAL	107.17	-20.80	94.68	-3.28	-2.49	26.38	102.63	65
45	MUSTEK SYSTEMS INC.	100.68	88.68	67.13	13.17	42.98	20.83	35.97	590
46	MICROELECTRONICS TECHNOLOGY INC.	99.17	1.54	170.38	7.89	132.23	72.60	22.37	976
47	DIAMOND FLOWER ELECTRIC INSTRUMENT CO., LTD.	98.60	23.54	42.45	0.83	25.40	N/A	40.17	305
48	CMC MAGNETICS CO., LTD.	96.00	-0.81	218.34	8.45	134.42	87.96	38.43	810
49	TRANSCEND INFORMATION INC.	89.81	61.79	41.40	3.70	13.62	6.11	67.09	88
50	LOGITECH FAR EAST LTD.	88.68	-6.15	47.36	-1.96	28.45	9.06	39.92	460
51	TAIWAN TELECOMMUNICATION INDUSTRY CO., LTD.	86.94	34.65	76.68	2.98	25.32	21.66	66.97	722
52	WNITECH PRINTED CIRCUIT & BOARD CORP.	84.94	62.88	73.17	10.23	34.53	22.64	52.75	696
53	ACER LABORATORIES INC.	83.47	30.81	52.00	1.92	9.13	6.64	82.43	205
54	TECOM CO., LTD.	80.72	11.34	99.43	4.53	70.42	44.87	29.18	423
55	KYE SYSTEMS CORP.	79.89	45.89	47.17	5.36	25.21	19.25	46.56	422
56	CHAINTECH COMPUTER CO., LTD.	78.53	33.82	42.00	2.45	15.92	12.08	62.08	256
57	ACCTON TECHNOLOGY CORP.,	77.40	35.64	62.04	8.00	39.66	21.09	36.07	537
58	HOLCO ENTERPRISE CO., LTD.	73.58	17.89	39.02	N/A	6.04	4.45	84.52	75
59	FEATRON THCHNOLOGIES CORP.	72.94	15.47	N/A	N/A	N/A	13.66	-	320
60	BIOSTAR MICROTECH INTL CORP.	72.26	-	32.60	N/A	8.75	3.77	73.14	106
61	VERIDATA ELECTRONICS INC.	71.25	101.70	63.77	-3.74	10.75	N/A	83.13	331
62	LUNG HWA ELECTRONICS CO., LTD.	65.62	27.86	37.74	1.36	17.62	16.04	53.30	200

Rank	Company	1995 Sales	Sales growth (%)	Assets	Profit after taxes	Net worth	Capital	Debt/Assets %	Employee Number
63	BRIDGE INFORMAION CO., LTD.	64.64	-7.65	37.81	0.49	8.91	6.38	76.44	245
64	PRESIDENT TECHNOLOGY INC.	63.89	58.66	65.32	N/A	-5.02	16.94	107.68	78
65	ABIT COMPUTER CORPORATION	61.13	14.32	32.34	0.49	5.21	3.77	83.89	147
66	MEGAMEDIA CORP.	60.57	-34.30	134.42	4.91	66.83	47.43	50.25	838
67	TAIWAN MYCOMP CO., LTD.	58.38	-11.14	39.25	N/A	11.92	7.55	69.61	156
68	FUNAI ELECTRIC CO. OF TAIWAN	54.38	-2.17	13.40	-0.83	5.09	7.92	61.97	320
69	V-TEC TECHNOLOGY (TAIWAN) INC.	53.40	27.13	15.47	0.30	1.32	0.75	91.46	32
70	ATEN INTERNATIONAL CO., LTD.	52.91	42.19	27.28	0.60	7.58	5.70	72.19	180
71	TRACE STORAGE TECHNOLOGY CORP.	52.38	173.76	85.55	6.57	23.89	N/A	72.07	550
72	LEO SYSTEMS INC.	50.38	390.80	61.58	-1.96	26.68	26.00	56.67	591
73	A-TREND TECHNOLOGY CO., LTD.	49.43	45.07	46.53	N/A	8.00	6.04	82.88	90
74	ADVANCED SCIENTIFIC CORP.	48.98	19.96	56.94	3.06	32.79	30.19	42.41	260
75	SUNRES TECHNOLOGY CORP.	48.30	18.29	28.11	0.75	15.58	15.02	44.43	139
76	TAICOM DATA SYSTEMS CO., LTD.	46.57	83.63	38.53	0.49	16.38	15.09	57.49	251
77	ZYXEL COMMUNICATIONS CORPORATION	45.06	8.44	55.58	12.87	46.75	24.79	15.88	237
78	LANDIS & GYR TAIWAN LTD.	44.72	21.29	37.13	-0.11	2.53	1.13	93.29	128
79	UNITRON INC.	44.45	3.15	31.17	0.60	14.72	11.32	52.78	85
80	MONTEREY INTERNATIONAL CORP.	42.45	-10.14	31.55	3.02	17.43	11.92	44.73	80
81	LUCKY STAR TECHNOLOGY CO., LTD.	41.62	53.19	20.49	N/A	1.58	1.09	92.26	65
82	CNET TECHNOLOGY INC.	41.51	44.73	28.38	3.06	19.28	15.32	31.91	165
83	LOYALTY FOUNDER ENTERPRISE CO., LTD.	40.57	48.89	40.49	2.15	12.83	3.77	68.31	220

Rank	Company	1995 Sales	Sales growth (%)	Assets	Profit after taxes	Net worth	Capital	Debt/Assets %	Employee Number
84	ENLIGHT CORP.	39.21	23.98	39.43	5.02	20.53	N/A	47.94	199
85	MACASE INDUSTRIAL CORP.	36.60	29.85	17.74	N/A	4.38	1.89	75.31	91
86	TXC CORP.	36.08	-19.59	37.17	-1.02	21.36	20.98	42.53	375
87	SYSGRATION LTD.	34.19	4.01	34.72	6.45	29.09	20.34	16.19	422
88	ADVANTECH CO., LTD.	34.11	45.57	17.55	1.02	8.57	7.17	51.18	220
89	GORDIA TECHNOLOGY CO., LTD.	34.00	-19.76	6.04	0.08	1.28	0.75	78.75	34
90	PRINCETON TECHNOLOGY CORP.	33.96	7.01	21.96	0.75	10.57	7.55	51.89	100
91	LEOCO CORP.	32.98	15.60	26.68	1.13	7.36	4.98	72.41	435
92	KUNNAN ENTERPRISE LTD.	32.26	-72.74	121.47	-108.79	-192.94	85.40	258.83	724
93	DAH YANG INDUSTRY CO., LTD.	30.45	17.63	32.19	N/A	15.70	8.68	51.23	230
94	RITEK INCORPORATION CO., LTD.	30.34	33.77	61.28	5.43	30.79	19.66	49.75	252
95	GREAT TEK CORP.	30.19	-14.16	27.25	N/A	2.00	1.70	92.65	100
96	FLEXUS COMPUTER TECHNOLOGY INC.	30.00	26.99	8.23	0.23	1.47	0.83	82.11	37
97	PHIHONG ENTERPRISE CO., LTD..	29.55	10.12	42.60	0.64	8.34	8.64	80.42	251
98	TAILYN COMMUNICATION CO., LTD.	29.36	72.50	43.13	2.26	23.32	13.28	45.93	175
99	SOUTHERN INFORMATION SYSTEM INC.	26.98	12.42	57.43	-1.17	22.08	16.75	61.56	125
100	GOOD MIND INDUSTRIES CO., LTD.	25.89	-24.28	10.79	-0.75	2.19	1.36	79.72	230

Appendix D: Taiwan's Major Information Electronics and Communication Products Manufacturers

Company	Founded	Location	President	Contact	Phone	Fax	1	2	3	4	5	6	7	8	9	10	11
A Plus Info Corp.	1990	Taipei	Dennis Wu	Dennis Wu	02-917-4591	02-911-0240	x										
A-Four Tech Co., Ltd	1987	Taipei	Robert Chen	Teresa Tsai	02-218-4552	02-218-9908						x					
A-Trend Technology Co., Ltd.	1991	Taipei	Eric Wu	Jeff Tsai	02-783-1125	02-785-5790	x										
Accton Technology Corp.	1988	HSIP	S. King	A.T. Huang	035-770270	035-770267								x			x
Acer Inc.	1976	Taipei	S. Shih	Simon Lin	02-545-5288	02-545-5308	x	x		x	x		x				
Acer Laboratories, Inc.	1986	Taipei	C.J. Wu	J.K. Pen	02-545-1588	02-719-8691								x			
Acer Peripherals, Inc.	1984	Taoyuan	K.Y. Lee	Hermit Huang	03-329-4141	03-329-3922			x						x		
Acer Sertek, Inc.	1976	Taipei	Jeff Chen	Phidias Chou	02-501-0055	02-502-8549						x					
Action Electronics Co., Ltd.	1988	Chung Li	J.P. Peng	Barry Chang	03-451-5494	03-452-0697	x						x				
Actown Corp.	1990	Taipei	Sunny Wang	Arthur Mar	02-218-0598	02-218-0599					x						
Adda Technologies, Inc.	1990	Taipei	Michael Kuo	David Wang	02-226-3630	02-221-4538	x						x				
ADI Corp.	1979	Taipei	James Liao	Donald Yang	02-713-3337	02-713-6555			x				x				
Advanced Scientific Corp.	1989	HSIP	Joseph Chang	Joseph Chang	035-780261	035-782868	x		x	x							
Advanced Video & Audio Tech.	1993	Taipei	T.H. Kao	C.L. Chiu	02-501-4126	02-504-3347							x				
Alphacom Enterprise Co. Ltd	1982	Taoyuan	Jerry Chen	Jerry Chen	03-480-1200	03-470-9313	x										
Aquarius Systems Inc.	1983	Taipei	Kenny Wu	H.Y. Liu	02-788-8066	02-788-7234	x			x							
Arche Technologies Inc.	1968	Taipei	Kunnan Lo	Philip Lin	02-556-4393	02-556-7192	x			x		x					
Artdex Computer Corp.	1989	Taipei	Mike Kao	Shelly Wang	02-634-6050	02-642-0528			x		x						
Askey Computer Corp.	1989	Taipei	Robert Lin	Alan Kao	02-218-3849	02-218-2845									x		
Aten International Co., Ltd.	1981	Taipei	C. T. Chen	U. Yang	02-504-7270	02-504-7282	x	x									
ATL Inc.	1958	Taoyuan	T.R. Lin	H.I. Yang	03-452-4141	03-452-9107	x	x					x				

Company	Founded	Location	President	Contact	Phone	Fax	1	2	3	4	5	6	7	8	9	10	11
Avision, Inc.	1990	HSIP	Thomas Sheng	Keun Hwang	035-782388	035-777017						x					
Axion Technology Co., Ltd.	1990	Taipei	Chuck Li	Yue-Je Yang	02-917-4550	02-917-3200							x				
Billion Electric Co., Ltd	1973	Taipei	C.F. Chen	Tim Chen	02-914-5665	02-918-6731	x										
Bridge Information Co., Ltd	1990	Taipei	Milton Chen	Willy Wei	02-298-8345	02-298-8385						x					
Cameo Communication Inc.	1991	HSIP	Y.T. Wen	Paul Jeng	035-779288	035-777461											x
Cheer Electronics Corp.	1982	Taipei	L.R. Guang	Robert Ho	02-225-9686	02-225-9676						x	x				
Chaintech Computer Co., Ltd.	1986	Taipei	R. Tung	R. Wu	02-240-7000	02-248-3009	x							x			
Chen-Source Inc.	1986	Taipei	Maggi Chen	Maggi Chen	02-248-9505	02-240-0965	x					x	x				
Chenbro Micom Co., Ltd.	1983	Taipei	Frank Chen	Maggi Chen	02-248-9505	02-240-0926	x										
Cheng Hong Chemical Co., Ltd.	1960	Taipei	J.W. Chen	R. Chiu	02-778-7998	02-741-7128	x	x						x			
Chic Technology Corp.	1990	Taipei	Samuel Tseng	Wendy Chiang	02-245-2737	02-245-2725	x										
Chicony Electronics Co., Ltd.	1983	Taipei	Kent Hsu	Charles Bao	02-298-8120	02-298-8442					x						x
Chien Hou Electronics Co., Ltd	1982	Taipei	H.T. Chi	Joanna Yang	02-725-2981	02-725-1993			x				x				
Chilong Data Products Corp.	1978	Taipei	M.C. Wei	Patrick Mii	02-773-8800	02-781-7990											x
Chin Ta Ind. Co., Ltd.	1971	Taipei	C.L. Chu	Mel Teng	02-202-1234	02-202-1230	x										
Chroma Ate, Inc.	1984	Taipei	Leo Huang	Irene Yuan	02-298-3855	02-298-3596			x								
Chun Yun Electronics Co., Ltd	1984	Taipei	K.Y. Lim	K.Y. Lim	02-992-6363	02-991-8483						x	x				
Chung Yu Electronics Co., Ltd.	1975	Taipei	J. Huang	J. Huang	02-365-0746	02-368-1968	x			x							
Chuntex Electronic Co., Ltd.	1981	Taipei	Frank Liu	H.P. Shen	02-917-5055	02-917-2736						x	x				
CIS Technology Inc.	1984	HSIP	Jack Chang	Vincent Kuan	02-918-7099	02-918-7145	x										
Clevo Co.	1983	Taipei	Kent Hsu	Y.T. Lee	02-299-1368	02-299-1360					x						
Cnet Technology Inc.	1989	HSIP	John Hsuan	Eddie Tseng	035-785158	035-782458									x	x	x

Company	Founded	Location	President	Contact	Phone	Fax	*Product Category 1	2	3	4	5	6	7	8	9	10	11
Compal Electronics Inc.	1984	Taipei	S.H. Hsu	Frank Yu	02-746-8446	02-760-7903					x		x				
Compro Technology Inc.	1988	Taipei	Susan Hsu	Andy Cheng	02-918-0169	02-912-1465		x						x			
Costar Electronics, Inc.	1987	Taipei	Victor Sheu	Jerry Pan	02-883-4633	02-881-0594									x	x	
CT Continental Corp.	1988	Taipei	Tina Kao	Peter Lin	02-627-0001	02-799-7691	x	x				x			x		x
CY&S Industrial Co., Ltd	1986	Taipei	James Van	James Van	02-296-4273	02-909-6742			x								
Data System Technology Co.	1985	Taipei	Julie Wang	Julie Wang	02-325-0851	02-325-2175			x						x		x
Datatech Enterprises Co., Ltd	1982	Taoyuan	Duke Liao	Simon Ho	03-328-3801	03-328-4058	x	x		x	x		x	x	x		
Demotek Computer Inc.	1991	Taipei	T.C. Deng	Michael Lin	02-833-1361	02-833-8276	x							x			
DTC Technology Corp.	1986	Taipei	David Lee	Dennis Ting	02-218-3880	02-218-9885								x		x	
Dual Group	1981	Taipei	Frank Tu	Victor Huang	02-788-3919	02-783-0023	x				x						
E-San Electronic Co., Ltd	1982	Taipei	Frank Chen	David Lien	02-755-6788	02-705-2547	x	x		x	x	x		x		x	
Edimax Technology Co., Ltd	1986	Taipei	Marcell Jean	Hausou Hsieh	02-299-5648	02-299-5647				x	x				x		x
Emperor Corp.	1983	Taipei	James Chao	James Chao	02-221-7155	02-223-6349	x					x					
Essex Monitor Co., Ltd	1989	Taipei	Rowell Yang	Joseph Yang	02-231-6789	02-231-5678						x					
ETC Group Fair Elec. Co., Ltd.	1983	Taipei	Jack Wang	Ocean Tseng	02-298-1480	02-298-1131				x		x					
Ether Electronics Co., Ltd.	1987	Taipei	David Lin	David Mao	02-504-5898	02-504-4611						x					
Ever Case Technology, Inc.	1984	Taipei	Daniel Yeh	Rita Hu	02-796-3049	02-796-2548	x										
Extech Enterprise Co., Ltd.	1986	Taipei	Ginger Tai	Ginger Tai	02-623-7121	02-722-7465	x										
Fast Fame Computer Co., Ltd.	1986	Taipei	Rolder Shiao	R. Shiao	02-286-6447	02-287-9051	x						x				
Featron Technologies Corp.	1993	Taipei	P.L. Hsu	John Fan-Chiang	02-226-0168	02-226-0188					x						
Feng Chi Trading Co., Ltd.	1961	Taipei	C.W. Tsai	C.W. Tsai	02-314-1226	02-361-3561						x					
First International Computer, Inc.	1980	Taipei	Ming Chien	Peter Ou	02-717-4500	02-718-2782	x	x		x	x						x

Company	Founded	Location	President	Contact	Phone	Fax	1	2	3	4	5	6	7	8	9	10	11
Fong Kai Industrial Co., Ltd.	1985	Taipei	Tony chang	Sabriwa Yen	02-299-3990	02-299-3752	x							x	x		
Forefront International Ltd.	1976	Taipei	Y. K. San	Lisa Twu	02-715-0457	02-713-0417				x		x					
Formosa Industrial Computing Inc.	1984	Taipei	Sharming Lin	Sharming Lin	02-226-4627	02-226-0840		x				x					
Fortuna Electronic Corp.	1974	Taipei	C. F. Ho	Mike Hsu	02-531-6366	02-537-5275	x			x	x	x					
Freetech Corp.	1986	Taipei	Annie Lin	Daniel Wang	02-720-5418	02-722-2746						x	x			x	
Full Yes Industrial Corp.	1987	Taipei	J. Chou	Grace Chiu	02-917-6633	02-917-1155	x	x									
Giga-Byte Technology Co., Ltd.	1986	Taipei	Dandy Yeh	Jack Ko	02-918-4839	02-918-4842	x										
GIT Co., Ltd.	1983	Taipei	Michael Hou	Steven Ho	02-248-9908	02-240-4256	x										
Good Way Industrial Co., Ltd.	1982	Taipei	Robert Tsao	R. Tsao	02-746-8270	02-761-2031	x	x						x		x	
Grand Computer Corp.	1983	Taipei	Jonny Wei	Sandy Liu	02-785-9197	02-785-5724	x							x		x	
Great Computer Corp.	1989	Taipei	Jim Lai	Jeff Lee	02-694-6687	02-694-6875									x		
Greentronix Inc.	1993	Taipei	S. Chen	K. Kuo	02-299-3456	02-299-2005											
Holco Enterprise Co., Ltd.	1983	Taipei	David Yu	Al Su	02-797-9668	02-799-7073	x	x					x				
Hopax Industries Co. Ltd.	1978	Taipei	Mr. Ho	Hars Chang	02-351-8095	02-351-7055		x									
Howteh Enterprise Co., Ltd.	1973	Taipei	John Chen	Mark Yang	02-771-6333	02-721-8513	x							x		x	
Hsing Chau Industrial Co. Ltd.	1975	Taipei	Gordon Su	Peter Wu	02-735-2167	02-732-3023								x		x	
Imtek Computer Co., Ltd.	1990	Taipei	Alan Tan	Alan Tan	02-723-8563	02-723-2541	x	x					x				
Intercom Technology Co., Ltd.	1987	Taipei	Victor Chang	Alfred chun	02-783-9943	02-783-5523	x				x			x	x	x	
Intracom Asia Co. Ltd.	1987	Taipei	Michael Thiel	Klavs Kruppa	02-723-1775	02-723-1776			x				x	x	x		
Inventec Corp.	1975	Taipei	Kou-I Yeh	Fred Chang	02-881-0721	02-881-0234	x				x						
Itaico Co., Ltd.	1990	Taipei	P. Zambon	P. Zambon	02-551-6248	02-551-6201	x	x	x	x		x	x				
Jackwood Enterprise Co. Ltd.	1971	Taipei	James Chen	Albert Tchen	02-765-4586	02-765-4601	x		x				x				

Company	Founded	Location	President	Contact	Phone	Fax	Product Category* 1	2	3	4	5	6	7	8	9	10	11
Jazz Hipster Corp.	1981	Taipei	Jazz Hsu	Ted Kao	02-222-5678	02-223-1333	x										
Jean Co., Ltd.	1986	Taipei	Jean Wen	Charles Kuo	02-718-7000	02-718-7171	x	x					x				
Jetta Computer Co., Ltd.	1989	Taipei	B.J. Wang	B.J. Wang	02-299-1750	02-299-1751	x				x				x		
Jump Cheer Enterprise Co., Ltd.	1988	Taipei	Anderson Liu	Taylor Huang	02-266-0050	02-262-0014	x										
Jung Mao Electronic Co., Ltd.	1978	Taipei	Albert Wang	Albert Wang	02-982-6350	02-986-4934					x			x	x		
K&C Technologies, Inc.	1982	Taipei	T.L. Hsu	Michael Mai	02-736-8958	02-737-2072			x					x			x
Kaimei Electronic Corp.	1973	Taipei	C.H. Chang	C.H. Chou	02-776-5922	02-752-8086	x	x		x							
Kalegen Enterprise Corp.	1988	Taipei	Cyndi Tung	Cyndi Tung	02-999-6118	02-999-6117						x	x				
Kapok Computer Co.	1992	Taipei	Simon Tsai	Steven Wang	02-298-2651	02-298-1128					x						
Kentex Electronic Co., Ltd.	1982	Taipei	T.J. Shu	Stefan Yen	02-221-5252	02-222-5948	x										
Kouwell Electronics Corp.	1980	Taipei	J.H. Tang	Tracy Chu	02-783-1166	02-783-5500								x			
Kuo Feng Corp.	1966	Taipei	Lin Shur Pu	Peter	02-754-8353	02-754-2829						x	x				
Kye Systems Corp.	1983	Taipei	Albert Chen	Jackie Hwang	02-995-6645	02-995-6649	x				x					x	
Lantech Computer Co.	1986	Taipei	Thomas Lee	Thomas Lee	02-766-7088	02-766-6892								x		x	
Lapro Corp.	1988	Taipei	C.S. Hsueh	Susanna Chang	02-226-4538	02-226-4238				x							
Leadtek Research Inc.	1986	Taipei	K.S. Lu	Kenny Chu	02-248-4101	02-248-4103	x										
Loyalty Founder Enterprise Co., Ltd.	N/A	Taoyuan	L.S. Chang	Jimmy Jane	03-326-9123	03-326-9111	x										
Lucky Star Technology Co., Ltd.	1983	Taipei	M. Huang	Jackson Wang	02-299-0222	02-299-0112	x			x				x			x
Lutron electronic Enterprise Co., Ltd.	1976	Taipei	D.C. Lin	Dale Tsai	02-596-0796	02-594-4901			x								
Lxycon Computer Co., Ltd.	1980	Taipei	Y.S. Teng	Alex Liu	02-298-3970	02-298-8231	x										
Marson Technology Co., Ltd.	1990	Taipei	Kenneth Liou	Perry Ho	02-785-7475	02-785-7483					x						
MEC Imex Inc.	1976	Taipei	Steve Hou	Alan Wang	02-772-2594	02-721-7175							x				

Company	Founded	Location	President	Contact	Phone	Fax	1	2	3	4	5	6	7	8	9	10	11
Microstar Computer Corp.	1987	Taipei	Chen In	Steven Wang	02-299-2229	02-299-2230	x				x						
Mighty Exim Corp.	1972	Taipei	Frank Huang	Teddy Yeh	02-753-3535	02-753-3237	x					x			x		
Ming Fortune Industry Co.	1983	Taipei	Wins Yang	Ms. Chiu	02-918-6079	02-918-6084			x								x
Mitac Inc.	1974	Taipei	N/A	Larry Yu	02-501-2650	02-509-0979	x			x			x				
Mitac International Corp.	1982	Taipei	Francis Tsai	Billy Ho	02-501-8231	02-501-4265	x	x		x					x		
Mr. Info Co., Ltd.	1987	Taipei	J.Y. Lo	Oliver Wang	02-221-1284	02-222-9469	x						x				
Mustek Corp.	1984	Taipei	M.Z.Chen	Bruce Tsao	02-882-6277	02-881-9970	x	x		x		x			x	x	x
Myday Technology	1992	Taipei	Duncan Kuo	Philip Wang	02-918-9211	02-918-9012			x								
National Datacomm Corp.	1989	HSIP	W.H. You	John Lin	035-783966	035-777989									x	x	
Northman Technologies Co.	1990	Taipei	Frank Tseng	Joanna Wu	02-788-4441	02-783-8913				x	x		x				x
Oryx International	1992	Taipei	Roger Lin	Angela Chen	02-935-6251	02-934-3274									x		
Paoku P&C Co., Ltd.	1972	Taipei	Rong Ling	W. Hsiang	02-391-4502	02-391-5359	x	x		x	x		x				
PE-Von Co., Ltd.	1989	Taipei	Judy Chang	Judy Chang	02-779-2188	02-779-0433		x		x							
Pei Chow Industry Co., Ltd.	1977	Taipei	Michael Hsu	M. Hsu	02-895-0556	02-896-2882						x					
Philips Electronics Ind. (Taiwan)	1977	Taoyuan	Y.C. Lo	Ares Hu	03-451-1308	03-462-0629						x	x				
Pinnacle Technologies	1984	Taipei	Bob Niu	Bob Niu	02-788-4800	02-651-2307			x				x				
Plustek Inc.	1986	Taipei	Karen Ku	Karen Ku	02-720-4101	02-720-8686					x						
President Technology Corp.	1967	Taoyuan	John Cheng	Alex Chang	03-482-9000	03-482-1278	x	x		x		x					
Primax Electronics Ltd.	1984	Taipei	Raymond Liang	H.C. Ho	02-695-3073	02-695-7064						x					
Princeton Technology Corp.	1986	Taipei	Richard Chiang	Jason Lee	02-917-8856	02-917-3836	x							x	x		
Princo	1983	HSIP	P.L. Chiu	S.C. Wang	035-773177	035-777456	x										
Pro-Ntes Technology Corp.	1993	Taipei	C.J. Wu	Wes Lin	02-916-1235	02-916-6488	x	x						x	x		

Company	Founded	Location	President	Contact	Phone	Fax	Product Category*										
							1	2	3	4	5	6	7	8	9	10	11
Prodem Technology, Inc.	1986	Chungli	Michael Huang	Grace Yueh	03-457-7314	03-457-7418		x									
Protech Systems Co., Ltd.	1980	Taipei	Engel Wu	Jesse Chiou	02-786-3173	02-786-2254	x			x	x			x			x
Puretek Industrial Co., Ltd.	1986	Taipei	Sheery Hsien	Sherry Hsien	02-916-1001	02-916-1004	x	x							x		
Rayon Technology Co., Ltd.	1990	Taipei	Mike Wang	Mike Wang	02-694-9460	02-694-9207			x				x	x			x
Regent Technologies Corp.	1993	Taipei	Ed Lee	Eva Loo	02-918-5123	02-918-5121							x				
Saho Corp.	1976	Taipei	Gibb Yang	Rual Lee	02-594-5011	02-594-4776							x				
Sampo Technology Corp.	1976	Taoyuan	Jung C. Ko	Rick Teng	03-328-5523	03-328-5529							x		x		
Shing Yunn Electronics Enterprise	1983	Taipei	Aaron Hsu	David Ren	02-203-6130	02-203-6125	x			x							
Samtech Corp.	1984	Taipei	Sam Lee	Sam Lee	02-214-2626	02-214-0217					x						
Shamrock Technology Co., Ltd.	1990	Taipei	S.C. Liang	Chris Lee	02-218-2155	02-218-5154						x					
SMT Electronics Co., Ltd.	1989	Taoyuan	Eric Chang	Carina Ong	03-326-1311	03-326-1142	x			x							
Snobol Industrial Corp.	1986	Taipei	Robert Chen	Varun M.C.	02-881-0991	02-881-1511	x							x			
Southern Information System Inc.	1981	Taipei	Jason Liu	Jason Liu	02-752-7006	02-751-2897			x								x
Soyo Technology Co., Ltd.	1985	Taipei	P. Sang	Sashia Lee	02-767-2115	02-766-3318				x				x			
Spring Circle Computer Inc.	1982	Taipei	Daniel Shih	D. Shih	02-918-4843	02-917-0551				x							
Summit Computer Tech. Co.	1988	Taipei	H.K. Lee	John Clark	02-906-3811	02-903-0051		x									
Sunshine Merchandise Promotion Co., Ltd.	1981	Taipei	A. Huang	Roger Tseng	02-501-5157	02-501-3899	x			x							
Super Elite Technology Co., Ltd.	1983	Tainan	C.N. Sun	M.D. Chen	06-269-8521	06-269-6598		x	x								
Surecom Technology Corp.	1987	Taipei	Mike Hsieh	Lilian Ou	02-592-2327	02-591-2675										x	
Systex Corp.	1976	Taipei	Paul Kao	Anna Lee	02-356-1277	02-356-8422			x								
Tailyn Communication Co.	1980	Taipei	Bruce Chen	Hank Chen	02-771-8916	02-741-1672	x		x								
Taimax Taiwan Computer Co., Ltd.	1992	Taipei	Mark Lin	Chung Bao	02-298-0298	02-298-0306						x					

Company	Founded	Location	President	Contact	Phone	Fax	1	2	3	4	5	6	7	8	9	10	11
Taiwan Mycomp Co., Ltd	1983	Taipei	M.C. Liu	Jack Liang	02-782-0201	02-782-7486	x						x				
Taiwan Semi Conductor Co., Ltd.	1978	Taipei	Arthur Wang	Corbin Fong	02-917-4145	02-912-2499										x	
Taiwan Turbo Technology Co.	1988	Taipei	John Chen	Pelly Liu	02-240-4918	02-248-8314	x	x									
Taiwan Video & Monitor Corp.	1978	Taipei	Michael Chiang	Nick Lee	02-776-5318	02-721-4798						x					
Taiwan Video System Co., Ltd.	1982	Taipei	Jeffrey Chen	Ami Chang	02-648-5524	02-648-5684						x					
Tamarack Microelectronics Inc.	1987	Taipei	P. Huang	Francie Liu	02-772-7400	02-776-0545								x			x
Tamarack Telecom Inc.	1990	Taipei	Carlai Ma	Steve Chang	02-356-9865	02-356-9868					x						
Targetek Inc.	1992	Chung Li	Ms. V. Yen	Richard Yeh	03-462-2111	03-462-2200	x	x	x	x	x			x	x		x
Teco Information Systems Co.	1989	Taoyuan	Eton Huang	Abraham Leu	03-473-3111	03-473-3126					x	x					
Tera Electronic Co., Ltd.	1976	Taipei	M.T. Hong	Hilton Chiang	02-882-3410	02-882-5465				x		x	x	x			
TFL LAN Inc.	1980	Taipei	Evergreen Lee	Kent Lin	02-882-5641	02-883-0405							x				x
Tobishi Electronic Co., Ltd.	1965	Taipei	C.S. Chen	C.S. Chang	02-202-0111	02-204-0656		x				x					
Transcend Information Inc.	1988	Taipei	C.W. Shu	Kathy Hou	02-788-1000	02-788-1919				x	x			x	x		
Tremon Enterprises Co., Ltd.	1985	Taipei	Steve Yang	Steve Yang	02-506-9074	02-507-3912			x								
Truedox Technology Corp.	1987	Taipei	Jaff Lee	Gloria Tu	02-222-6196	02-222-6204	x					x					
Twinhead International Corp.	1984	Taipei	Stanley Chiang	John Chin	02-917-9036	02-917-2675				x	x						
TXC Corp.	1983	Taipei	William Hsu	Lawrence Ko	02-894-1202	02-894-1206	x				x			x			
Uerro Industrial Co., Ltd.	1977	Taipei	Liao Yi Ming	Eric	02-953-2662	02-955-5607				x	x	x					
Umax Data System Inc.	1987	Taipei	Frank Huang	Jack Lee	02-517-0055	02-517-2017	x					x					
United Hitech Corp.	1989	Taipei	Frank Lee	James Lin	02-299-3125	02-299-3210	x										
United Technology Corp.	1988	Taipei	Jame Yuan	Nelson Yu	02-501-5330	02-501-5914	x		x	x							
Unitron Incorporated	1979	Taipei	J. Chang	Rupert Cheng	02-218-1881	02-218-1398	x			x							

Company	Founded	Location	President	Contact	Phone	Fax	1	2	3	4	5	6	7	8	9	10	11
Universal Microelectronics Co.	1984	Taichung	Jimmy Ou	Jimmy Ou	04-359-0096	04-359-0129									x		
Unixtar Technology Inc.	1986	Taipei	Julian Hsien	Kyle Tseng	02-218-8863	02-218-2833			x						x		
Veridata Electronics Inc.	1988	Taipei	Robert Cheng	Jessie Lei	02-791-5490	02-791-8292				x							
Welltronix Co., Ltd.	1982	Taipei	F.Y. Lai	F.Y. Lai	02-838-2990	02-838-2992	x	x									
White Horse Industrial Co., Ltd.	1972	Taipei	L.T. Wang	Gary Shen	02-731-3988	02-731-3969		x							x		x
Win Technologies Co., Ltd.	1989	Taipei	Paul Van	Paul Van	02-788-9355	02-651-0315	x				x						
Win Way Co., Ltd.	1979	Taipei	Ellick Tsai	Adam Young	02-778-3195	02-741-9472						x					
Winny Electron Enterprise Co.	1986	Taoyuan	J.Y. Chang	Charly	03-457-4191	03-458-1267						x	x				
Wintech Electronics Corp.	1981	Taipei	J. Chang	Neil Hwang	02-923-9003	02-923-9023	x	x									
Wise Connection Corp.	1993	Taipei	Y.P. Lee	Jack Lin	02-688-5465	02-688-9027						x					
Wiso Electronics Co., Ltd.	1972	Taipei	Henry Tsai	Henry Tsai	02-393-6147	02-393-1469					x	x				x	
Yacase Industrial Corp.	1989	Taipei	Chad Lin	Fred Chen	02-523-0722	02-523-0647						x					
Yeong Yang Technology Co., Ltd.	1978	Taipei	Y.Y. Yen	Cathy Hsu	02-680-2339	02-680-2338		x									
Youth Keep Enterprises Co., Ltd.	1978	Taipei	K. Tsao	Karen Tsao	02-531-0170	02-564-2131	x	x						x			
Zebex Industries Inc.	1987	Taipei	Jerry Liao	Jerry Liao	02-218-3501	02-218-8670					x						
Zero One Technology Co., Ltd.	1980	Taipei	Peter Lin	Nick Huang	02-565-2323	02-561-9505										x	
Zyxel Communications Corp.	1987	HSIP	Shun-I Chu	Mark Hsia	035-783942	035-782439		x									

*Product Category: **1.** Main Circuit Boards, **2.** Multimedia Products, **3.** Fax/Modem, **4.** Desktop Computers, **5.** Portable Computers, **6.** Image Scanners, **7.** Video Display Devices, **8.** Interface Cards, **9.** PCMCIA, **10.** Printers/Plotters, **11.** LAN Equipments.